AND
BITES
THE STORY OF THE SILVER GLEN

Stephen Moreton

BONANZAS AND JACOBITES

THE STORY OF THE SILVER GLEN

Published on behalf of
Clackmannanshire Field Studies Society (CFSS)
with support from
Clackmannanshire Heritage Trust.

by NMS Enterprises Limited – Publishing
a division of NMS Enterprises Limited
National Museums Scotland
Chambers Street
Edinburgh EH1 1JF

10-digit ISBN 1-905267-08-8
13-digit ISBN 978-1-905267-08-8

British Library Cataloguing in Publication Data
A catalogue record of this book
is available from the British Library.

Cover design by Mark Blackadder.
Internal layout by NMSE – Publishing, NMS Enterprises Limited.
Printed and bound in Great Britain by Athenaeum Press Ltd, Gateshead, Tyne-on-Weir.

CONTENTS

Silver Glen, Alva.
The mines (not visible at this distance) are in the centre of the picture.

PREFACE

THE mines in the Silver Glen, east of Alva, near Stirling in central Scotland, have long been a local curiosity, but they receive surprisingly little attention in works on Scottish history or mineralogy despite having made significant contributions to both disciplines. A book by Bessie Dill published in 1909 and entitled *The Silver Glen, a story of the Rebellion of 1715* was, for a long time the only sizeable account of what went on at Alva House and the nearby mine in 1715. However, it is a fictional work based loosely around the letters of Lady Erskine. For many years the only published historical information was a few paragraphs in the Statistical Accounts and in the less well known *Between the Ochils and Forth* by David Beveridge (Wm Blackwood & Sons: Edinburgh, pp. 258-61), neither of which do justice to the subject. The journal of Henry Kalmeter, a Swede who visited the site in 1720, also makes a short reference to the mine. The English translation was published by the Scottish History Society in *Scottish Industrial History: a miscellany* in 1978. The Clackmannanshire Field Studies Society published a booklet, 'Mines and Minerals of the Ochils', in 1974 which drew on the first two sources and gave a brief description of the mine site – but still the full story remained untold.

It was not just silver that was extracted at Alva. The discovery of cobalt in 1758 led to much excitement at the time. Again, for a long time there was almost nothing published on this event until Jill Turnbull's account was written up in the *Transactions of the English Ceramic Circle* (1997, volume 16, part 2, pp. 144-51) – again much remained untold.

My own interest in Alva began in 1981 when, as a keen young mineral collector, I visited the Silver Glen hoping to find the bright pink mineral erythrite. After just three or four brief trips, I had found not just erythrite, but also the rare mineral clinosafflorite, together with the most exquisite crystallised native silver I had ever seen outside a museum.

Researches into the history of the mine were just as fruitful. Scattered throughout dozens of eighteenth-century hand-written documents, many barely legible, dispersed around record offices and libraries, was the story of the Silver Glen. It was an extraordinary tale of fortunes made and lost, of Jacobites and Hanoverians, of a loyal and loving wife, of buried treasure and sunken treasure, of treachery and betrayal, of conspiracy and greed, and with a cast as diverse as the Old Pretender and Sir Isaac Newton.

Rather than put it into my own words, I have let the characters themselves tell the story through their own writings as preserved in the archives. I have made no attempt to correct mistakes (spelling or otherwise) or to insert punctuation where, all too often, there is none. Since some of the people involved seem not to have known of the existence of the full stop, this can make reading their letters something of a challenge, but I did not wish to alter them in any way. I have also retained the imperial units. For those too young to remember, a fathom (6 feet) is approximately 183 cm, a ton (2240 pounds) about 1016 kg (just over a metric tonne), and a stone (14 pounds) about 6.4 kg. To confuse matters, one of the characters in the story used a stone of 16 pounds (7.3 kg). Precious metals were weighed by the troy ounce (31.1 grams) rather than the more familiar ounce avoirdupois (28.34 grams).

After more than 25 years my research is still not complete, but if I was to put off writing it up until I had seen every last Treasury receipt or mention of the mine, the story would never get told. More documents certainly exist. The papers of the Johnstone family, recently deposited in Alloa Library, are known to contain references to the mines. Unfortunately, the onerous task of cataloguing these has only just begun and could not be completed in time for publication of this book.

I have written it with both the general public and academics in mind. Those who only want to know the story can just read it and enjoy it. For those who want to see the sources, or who have a research interest, the book is thoroughly referenced, chapter by chapter, with additional notes and background information accompanying the references.

It would not have been possible to research and write this book without a great deal of assistance. I am indebted particularly to the National Archives of Scotland (formerly the Scottish Records Office), the National Archives (formerly the Public Records Office) and the National Library of Scotland for access to their historical manuscript collections and for permission to publish transcripts of these. I also thank Dr Ken MacKay of Cambusbarron for his help in locating documents and for some of the photographs and the Forestry Commission and the Woodland Trust (former and current owners respectively of the Silver Glen) for permission to collect minerals. Thanks are also due to National Museums Scotland and the Universities of Edinburgh and Manchester for mineralogical work. Jill Turnbull of Edinburgh provided information about cobalt and eighteenth-century porcelain. Brian Jackson (National Museums Scotland) and John Harrison of Stirling checked the manuscript and made helpful suggestions. Lindsay Corbett of the Clackmannanshire Field Studies Society provided contacts and information, and facilitated funding for the publication from the Clackamannanshire Heritage Trust, Lesley Taylor of NMS Enterprises Limited – Publishing has done an excellent job of turning my plain draft into a readable format, and others, too numerous to list, gave permission to use photographs or helped in other ways.

STEPHEN MORETON
2007

ABBREVIATIONS TO NOTES:

E-M	Erskine Murray
f	folio
LLC	London Lead Company
MSS	manuscripts
NLS	National Library of Scotland
NRS	Northumberland Records Office
NA	National Archives (Public Records Office), Kew, Surrey
NAS	National Archives of Scotland (Scottish Records Office)
WRH	West Register House

Sir John Erskine, from an old painting, present whereabouts unknown.
Picture taken from *The Silver Glen* by Bessie Dill, 1909.
(PHOTOGRAPH BY DR KEN MACKAY)

CHAPTER 1

THE DISCOVERY OF THE MINE AND THE 1715 REBELLION

WHEN Mexican silver miners discovered an exceptionally rich mass of ore, they would call it a *bonanza.* Such bonanzas frequently contained many tons of precious metal and often made their fortunate finders fabulously wealthy. Such deposits are not confined to Latin America, and the discovery of a small, but rich, silver bonanza in 1715 in the Ochil Hills near Stirling, coinciding with the Jacobite insurrection of that year, led to a remarkable and fascinating sequence of events.

The central character of the story was Sir John Erskine who lived at Alva House, about six miles east of Stirling. A distant descendant of Robert the Bruce, he was born in 1672, the second son of Sir Charles Erskine and his wife Dame Christian Dundas.[1] His elder brother, Sir James, was killed at the battle of Landen, Belgium, in 1693. His younger brothers included Dr Robert Erskine (1677-1718), physician to and favourite of the Russian Tsar Peter the Great, and Charles Erskine, Lord Tinwald, later Lord Justice-Clerk, who enters the story later. His many other siblings either died young or achieved no great eminence.

In 1705 he married the Honorable Catherine, second daughter of Lord Sinclair, a loyal Jacobite. They had three sons: Charles (killed at the battle of Laffeldt, Netherlands, in 1747), Henry and John. Although a landowner, Sir John's income was unlikely to have been great. In that period a Scottish landowner was considered wealthy if he had a rent roll of £500; rich if his income was £200-£300; and well-off with £80-£100 – while many gentlemen had to make do with even less. Commonly, much of the rental would be paid in kind; thus, in 1701, the Erskines' estate at Alva brought in a rental of £85. 10s. sterling, together with various quantities of oats, meal, corn, etc., some chickens, three geese, two lambs and a pig.[2]

Not surprisingly, landowners were always looking for ways to improve their estates and increase their incomes. Sir John was one of the first Scottish lairds to follow English examples and invest in enclosures and plantations, which were later to prove very worthwhile. One of his earliest ventures, however, was also one of the most speculative: metal mining. In those days mining for metals was very much a hit or miss affair. If the miners were lucky, rich deposits may be found and huge profits made. More often, trial mines would prove to be nothing more than expensive holes in the ground.

A wiser man may not have risked blowing what could amount to a year's rent on such a gamble, especially in an area with little history of mineral workings. (The largest local metal mine worked a small copper deposit at Bridge of Allan which, from time to time, made money but often lost it too.) But, as John Ramsay of Ochtertyre remarked of Sir John, 'He was a man of wit and genius, but the heat and volatility of his fancy would not be regulated by prudential considerations'.[3]

So it was that sometime between 1710 and 1714, Sir John brought in some miners from Leadhills to look for any promising indications of minerals on his estate at Alva. Their initial search seems to have been unsuccessful, but not their second attempt, in December 1714, as Sir John was to explain in a 'memorial' to Lord Viscount Townshend in 1716:[4]

> That in the Month of December One of these Miners made a Second Search for Mettals in the Mountain to the Westward of your Memorialist's House, and about Christmass found (what he called) a thread of Ore lying South and North, and with difficulty digged a Small quantity thereof, this Ore was tried by Sundry Persons, who made very different Reports concerning the Same only they seemed to agree that it was metal and worth the refining, so as it could easily be come at.
>
> During the rest of the Winter the Miner and one more were picking such Small pieces of that thread of Ore as the Hardness of the Rocks in which it lay would allow them. Till about the Month of Aprill your Memorialist sett two men more to Work to digg up what they could of this Ore about which time they found great variety of Ores, Some Rich

> in Silver, some Copper and some they Could make no Judgement of, neither give any Name to, never being able to reduce it to any kind of Mettal. The diggers continued at this Work some Weeks till the Ore began to be Scarce and that with the Hardness of the Rocks, and great Charge made your memorialist give over Working in that place and sett the Workmen to another Vein (about 3000 Paces farther in the Mountain) lying East and West, which had a very promising appearance, but after about Nine Weeks Work they Came to nothing like Metall and so gave it over.
>
> That in July your Memorialist was perswaded to make a farther Tyrall, where the Ore was found in April, and in some Weeks Work, the Vein appeared more promising but this happen'd a very little only before he Left his own house.

Sir John left his house because rebellion was in the air. On 20th August the Earl of Mar, a distant cousin of Sir John (and, confusingly, whose name was also John Erskine) arrived at Braemar to lead the rebellion in Scotland, and on 6th September the Jacobite standard was raised. An unsuccessful attempt on Edinburgh Castle two days later was followed by the seizure of Perth on the 14th by Colonel Hay, and of Inverness the same day by Mackintosh of Mackintosh. The assession of James III (of England) and VIII (of Scotland) was proclaimed at Aberdeen on 20th September and a violent confrontation between the Jacobites and the government supporters of George I was inevitable.

Shortly before Sir John left, one James Hamilton had appeared on the scene. Hamilton had been asked to come to work at the copper mine at Bridge of Allan by a Mr Daniel Peck, but on his arrival in August he found that the mine's owner had left to join the uprising. Peck then recommended Hamilton to Sir John who employed him to smelt the ore coming out of the newly-discovered mine. In eight days he extracted 433 ounces of silver, which Sir John took from him before his departure.[5]

Sir John seems to have spent most of his time in Europe, where he had been sent on 28th September to seek arms and ammunition, rather than fighting alongside the Earl of Mar. The Earl headed south to confront the government forces led by the Duke of Argyll. The rebel army outnumbered the Duke's forces of less than 4000 men by at least three to one. Their plan was to detach 3000 men to harass the Duke's army at Stirling, and, while the Duke was being kept busy, the main contingent would head for England.

It was not to be. The Duke's spies learned of the plan and his army marched to Sherriffmuir, about five miles north of Stirling. So it was that the Earl, who was not a good commander, was forced to do battle on ground of the enemy's choosing. The two armies clashed on Sunday the 13th of November. The churches were silent that day and the minister of Alva sent a message to Alva House that, 'there wad be nae Sabbath the day'. Local church records state: '1715, Nov. 13, no collection upon accoumpt of the Batell of Sheriffmuir.'

The battle was indecisive, and when the rebels heard that their comrades in England had been defeated that same day, at the Battle of Preston, things must have looked gloomy. Sir John, however, was safe in France, a fact which gave his wife much comfort for, in December, she wrote to him saying:[6]

> My Dearest Life,
>
> I receiv'd Yours of the 20th & another of the 29 of Nov which were both most acceptable, butt they had both been long by the way for itt was the 5 of De befor I receiv'd the first, You are much mistaken in thinking I was displeasd with you for leaveing this country I doe assure you I thought itt a lucky providence & tho I was in fear for not hearing from you yett itt was easy to bear in comparison of what terror I must have had if you had been in the danger some others of our freinds have been in, I suppose you know all our difficultys from better hands long ere now & by that you may guess the torment fear & terrible Horror I must be in for you & many others if I had known your adress I had writ to you three weeks agoe & begd of you to stay where you was till you saw how things wold be I writ to your Brother in hops he wold learn itt from some att Edr butt he told mee he could nott & you was soon expected & I was so far from wishing you soon back I was afraid to hear of your return, I pray God send a happy end to all for I am just where I was & my hops are still very faint. that person you mention in yours not being come yett your children are very

well & all your other freinds I doe nott wish to hear you are returned butt when you doe pray God you may be saffe which is the earsnest wish off her who is intirly

Yours

The letter was slow in reaching him; he did not receive it until the 15th February 1716, at St Germains. In the meantime, Lady Erskine had been keeping the miners busy at Alva. Under her orders James Hamilton supervised the labours of four miners. In the space of about three months they dug out as much ore as they could. The best of this was melted down and the remainder, about forty tons, was buried in barrels near the house.[4,5] All of this was conducted under the utmost secrecy; even the miners did not know what it was they were extracting, as is revealed by an interesting letter written by a servant of the Erskines, John James, to a George Allanson of Bristol about a year later:[7]

Sr Chepstew September ye 8th. 1716

I had ye favour of yours yesterday and according to your desier I have sent you account of what I know Relating to ye silver mine in north Brittain.

In ye year 1711 I became a servant to Sir John Arskine near Alva who Emplyed me to loock after his servants and improve his Estate after ye English maner I had not been thare long but he was often telling me he had a mine in the side of a hill near ye hous in which thare was a sort of ore that would pduce som metall of value if a pson of judgment mad tryall. whereupon I ofered my assistance and both he and I made severall Essays but without succiss for want of better knowlidge and meterials fitt for such an undertakeing however by ye small Experience found that a very good metell might be brought from it afterwards wee had ye asistance of a gouldsmith from allanay who prended to have great judgmt in reducing and refining metals but after many Essays left us both as much in ye darck as before so that ye undertaking lay ded for som time but afterward hearing of a very knowing man in London Sir John sent for him downe and wee made further inspection on the same ore but still in vaine not being able to bring ye metal soe as to forme a true judgement on ye value but at last trying all maner of wayes and things wee found out that which ansered our highest Expectations and ye ore so rich that from on pound weight coud make to ye value of 4^s and 4^d and upwardes sterling wee continued ye worck with so great success and secracy that servants belongin to the sam family never knew what we were about severall hundreds pounds value was Refined in a very short time. The ore was dug by three or fower of Sir Johns pore tenants that did not know what it wass In a short time after the work was begun sir john went to the pretender and never came to his ffamily afterwards and to secure all the plate then made from the kings troops my Lady ordered to be hid under the bordes in a upper flore and all the oar that was then dug was put into Casks and hid in the Earth to the value of many thousands of pounds but m^r hamilton nor my self was never trusted in that affair although wee both knew the place full well the min which which was dug is 4 . 5 and in some places 6 inches thick and Certainly is of many millians value. but I discovered one in an other mountain of about 12 inches and am allmost confident that no other person in the world besids my selfe knowes any thing of it but am redy and willing to discover the same the first opertunity to the Crowne I have brought a sample of the same ore with me to demonstrate to any pson ye government pleas to apoint that from on pound weight will mack 4^s:4^d true sterling and upward

M^r hamilton and selfe left scotland as soon as it was possible to get pass notwithstanding I left four years wages behind with a designe to mack a full discovery to the government of the whole matter as I hinted to you when last in Bristoll my whole dessigne as a returen to your former favours to me was to have m^r hamilton com downe to bristoll and take your advice how to pseed in this affair not douting but a discovery of soe highe a nature and advantag you might have som pffitt thereby last week I had an anser from m^r hamilton who writes me he was just going to Scotland and should be back in a short tim and this is the state of whole matter and for ye truth am willing to mack oath to the same if requird I desigin to wait on you in Bristoll in to or three dayes at most and from thence to London where m^r hamilton and I have a worrking Roome between us I have not to ad but am ass

your forever obliged humbell
Servant John James

While the miners were busy at Alva, Sir John was having his own fair share of excitement. Repeated attempts to procure arms for the rebels met with obstructions and difficulties, but he did manage to take charge of a valuable consignment of Spanish gold. Queen Mary, mother of the Pretender James, had received, according to various accounts, 100,000 ducats or 200,000 crowns worth of gold ingots from Philip V of Spain to assist the rebels. Sir John, together with Lord Tynemouth and Francis Bulkeley, was to convey this gold to Scotland where it was to be handed over to King James. Their ship set sail from Calais late in December and soon encountered stormy weather. Early in January 1716 it was driven onto a sandbank off St Andrews where it was smashed to pieces. Sir John and the crew survived but the treasure was lost. They lingered for a few days hoping to salvage something at low tide, but had to give up.[8]

On the 2nd of February 1716, Sir John was sent back to France by the Pretender and the Earl of Mar, with despatches to the Duke of Orleans, the Regent of France, the Pretender's wife and the Earl of Bolingbroke. He set sail from Montrose at midnight on the 3rd, the Pretender sailing from the same port a day later. Sir John arrived at Calais on the 6th and then made his way to St Germains where he delivered his letters to the queen on the 10th before leaving for Paris to look for Lord Bolingbroke.[9]

Sir John remained in France for several months and was frequently hard up for cash despite repeated attempts by his wife to make arrangements for £100 to be passed on to him. It seems there was some difficulty in communication as revealed by Lady Erskine's frequent letters to him. On April 13th she wrote:[10]

My Dearest Life,

I am most uneasy you have never got any of my letters & I am much afraid you are in want of money I have writ six letters since you left Britain & in

> every one of them begd you to cause your Factor Draw upon his Corespondant for 100 pound ....

and on the 4th of May:[11]

> ... this is the eight I have writ & tho by your last you tell me you had not heard from me I am hopfull they are not all miscaryed but by your leaveing Paris they ar longer a coming to your hand ....

In her letters she assured her husband that the estate was in good order, the hedges and ditches maintained and the crops sown. She also reported that their sons had 'Chincoch' (whooping cough), but were recovering. The one thing she carefully avoided any direct mention of is the mine, perhaps in case her letters should fall into the wrong hands. However, there are veiled references to it. Thus, on May 26th, 1716, she wrote, 'it is yet impossible to tell what money Mr Nabit will be worth his reputation amongst the comon sort is so high that no body credits it'.[12] At first this seems quite innocent, until one realises that 'Nabit' (now known as 'Nebit') is the name of the hill on which the mine was situated. Later, on 11th June, she wrote:[13]

> ... as to Mr Nabit his fame was too much spread abroad to contnue imploying old W & all of a sudden there was no more valuable thing to be got in that place & it was thought a lucky providence that it went quit away & so was shut up as to the value to the Stock in hand it is of so diffrent values that no mortall can make a just calculation as to DP his credit is quit gone & he owing a great many 100 pounds so that he dare nott appear ...

Presumably 'DP' was Daniel Peck who had introduced James Hamilton to the Erskines, and the 'Stock in hand' is probably a reference to the ore buried in barrels. 'Old W' refers back to her letter of 23rd March in which she wrote: '... as for Old W, work I am obliged to give it up yesterday untill we be in a state of more freedom than we are at present & people began to suspect there was something in it more than ordinary that I continued it so long ....' Perhaps 'Old W' was a miner.

It appears from these remarks that, despite all the secrecy, rumours were beginning to fly, and it was thought best to close up the mine. In fact, according to Sir John's memorial, the mine was abandoned and filled up with earth and stone towards the end of February.[4]

In July 1716 Sir John was sent to Sweden by the Pretender to attempt to enlist the aid of Charles XII, but he got no further than Lübeck. By this time the rebellion was in ruins anyway and the rebels either captured or, like Sir John, in exile. The whole affair seems to have been mismanaged from the beginning. The Earl of Mar was not the best choice of commander. What should have been a clear victory for the Jacobites at Sherriffmuir was ruined by indecision and missed opportunities.[14]

Sir John's main criticism was the lack of arms and ammunition. In a journal he kept,[9] he criticised those he believed responsible (particularly Lord Bolingbroke) and asked 'why was not ther some officers some ammunition & some arms sent in every little ship that went wch might ha made a sufficient provision of all needfull in due time?' There had been plenty of men willing to fight, he said, but not enough arms to go round.

Having been so heavily involved in the rebellion, Sir John was a wanted man. With the rebels being attainted and their estates confiscated his predicament was grim. Friends in London managed to delay action against Sir John for a while, but they could not put off the inevitable indefinitely, as is apparent from Lady Erskine's letter of 11th June:[13]

> ... you are not yet attainted so I hop will scape this session of Parlyment but if ever you are attainted all you could once call your own is irrecoverably lost there is such acts of Parlyment passing as people cannot expect to save any thing ...
>
> ... I am told by some you very narowly mist being putt in the last Bill of attainder & its affirm'd that your not being putt in was oweing Pet. H_ine if you still remain where you are at present its impossible you can scape being attainted as soon the Parlyment sits down ....

At first it was hoped that through the influence of Sir John's brother, Dr Robert Erskine, Tsar Peter the Great could assist in obtaining a pardon; but if this happened, Sir John would still be forbidden to return home.[11] In the meantime, Lady Erskine was preparing for the worst. On 18th June she wrote:[15]

> ... by degrees I am to sell all my stock & prepare for the worst I must own it was what I was mighty unwilling to doe but I am now convincet it's the best way by much, As to Mr Nabit I am sorry I have not writ so fully as you might understand his fame was like to rise High & at the same time there was never less ground for it ...

In the same letter she also mentioned, in passing, an event involving James Hamilton which was to have tremendous repercussions for the Erskines – although they did not know it yet: 'James went away three months agoe for he turnd wrong in the head & wold not stay.'

NOTES TO CHAPTER

1 Erskine family tree, NLS, E-M MSS, 5115.

2 Paul, R.: 'Alva House Two Hundred Years Ago: part II', in *The Hillfoots Record* (10th April 1901), p. 3.

3 Allardyce, A. (ed.): 'Scotland and Scotsmen in the Eighteenth Century' (Wm Blackwood & Sons: Edinburgh & London, 1888), volume II, p. 110.

4 NA, MINT 19/3/256-7, 9th October 1716.

5 NLS, Paul MSS, 5160, f. 5, 3rd July 1716.

6 NAS, GD1/44/7/13, 10th December 1715.

7 NA, T1/200/119-120, 8th September 1716.

8 The rebels tried to keep the affair of the gold secret, but without success, for the Hanoverian forces soon found the wreck and spent sometime on it 'fishing for gold'. See A. Tayler and H. Tayler: *1715: the story of the rising* (Thomas Nelson and Sons Ltd: London, 1936), pp. 144-145.

For further details and references, see also 'Spain and the Jacobites, 1715-16' by L. B. Smith in E. Cruikshanks (ed.): *Ideology and Conspiracy: Aspects of Jacobitism, 1689-1759* (John Donald Publishers Ltd: Edinburgh 1982), pp. 168, 176; and P. Miller: *James* (George

Allen & Unwin Ltd: London, 1971), pp. 195, 200: although the latter wrongly places the wreck at Dundee.

9 NLS, 5116, ff. 1-8, February 1716.

10 NAS, GD1/44/7/5, 13th April 1716.

11 Ibid., f. 7, 4th May 1716.

12 Ibid., f. 9, 26th May 1716.

13 Ibid., f. 10, 11th June 1716.

14 Tayler, A. and H. Tayler: *1715: the story of the rising* (1936), pp. 94-105.

15 NAS, GD1/44/7/12, 18th June 1716.

The Hon. Catherine, Lady Erskine,
from an old painting, present whereabouts unknown.
Picture taken from *The Silver Glen* by Bessie Dill, 1909.
(PHOTOGRAPH BY DR KEN MACKAY)

CHAPTER 2

THE GOVERNMENT STEPS IN

ON the 1st of March, 1716, James Hamilton left Alva and travelled to London, taking with him some samples of ore. Within days of his arrival he told the Lord Mayor of London, Sir Charles Peers, about the silver mine. Things seems to have moved slowly at first, as it was not until the 6th of June that Hamilton assayed some of the samples in the presence of Jeremiah Marlow and Peter Makin. A sample was also given to Sir Isaac Newton, Master of the Mint, who was impressed enough by his analyses to describe the ore as 'exceeding rich', a pound avoirdupois of ore producing 22 pennyweight of silver (i.e. 7.5% silver) worth about 5s. 7d.[1]

On the 3rd of July Hamilton gave the following formal affidavit to the Lord Mayor:[2]

> The Deposition of James Hamilton of the Parish of St Bridgets alias Brides London
>
> Came before me Sir Charles Peers Knight Lord Mayor of the city of London this third day of July 1716 and voluntarily made Oath that Mr Daniel Peck wrote this deponent to come down to him to Scotland to work upon Copper Mines belonging to one Dundas Laird of Maner and on the third of August last past went down on Horseback. When he came to Scotland he was informed by the Said Daniel Peck was gone into the rebellion, So nothing was done as to the working of the Copper Mines which are in a mountain called Ethry within two or three miles of Stirling. Upon which dissapointment Said Daniel Peck recommended this Deponent to Sir John Areskine who hired this Deponent to melt and refine his Silver Ore, on which he was imploy'd eight days in Said Sir John

Areskine's house and in that time refin'd from Said ore four hundred thirty three Ounces of virgin Silver, The which Sir John Areskine took from him to his own Custody And then Said Sir John Areskine went into the Rebellion with Six of his Servants and one of his Tennants, which Tennants name is John Marshall and now a Prisoner in Sterling Castle. Upon which Sir John Areskine's Lady then Imployed this deponent as overseer of four men that digg'd the Ore out of the Mine for above three Months in which time they dug out of the Said mine to the best of this Deponents Judgement and knowledge about forty tons of ore the which was brought to Sir John Areskine's house and ther pack'd in Pipes Hogsheads & other casks which they buried in a bank w[t]out the north-west end of the house Just without the Gate. Said Silver Mine is about a quarter of a Mile from Sir John Areskines house in the Parish of Alloa, within two miles of Alva a Seaport town in the water of forth. This Deponent further Saith that Sir John Areskin's Servants told this Deponent that their Master was returned from France with Arms and landed at St Andrews wher the Ship run on the Sands and was lost: and further Saith that Sir John Areskine returned again to France on Some Errand of the Pretenders as all the Servants of the house told this Deponent who tarried with the Lady Areskine at her house untill the first of March last past not daring to adventure to travel to London Sooner, when he forward to London and brought with him about eight or ten pounds of the Silver ore as came blasted out of the Rock. Promiscuously this Deponent further Saith that out of Sixteen Ounces of Avoirdupois of Said ore which he made ane Essay of produced above fifteen pennyweight and half of Virgin Silver w[c] Essay was made in presence of M[r] Jeremiah Marlow and Peter Makin on the Sixth day of June past by direction of the Right honble Sir Charles Peers Kt Lord Mayor of the City of London to whom I Communicated the above Contents few days after my Coming to London being well Informed of his Lordships great Zeal for his Sacred Majesty K: George and the Succession in his Royal line that he might reap Some Advantage by making it known to his Majesty and on Said Sixth day of June past brought Sealed up w[t] M[r] Jeremiah Marlow the Said piece of Virgin Silver and delivered

> into his Lordships own hands. And this Deponent further Saith that Sir John Areskine's Silver mine Veins the uppermost about twenty two inches thick to perfect Vein, and under that ane other about fourteen Inches thick and Contains to the best of this deponents Judegment above half a mile having view'd the Rock and found the Said ore at Said Distance. This Deponent further Saith that he was at Dundas Copper Mine and that it seems to be a noble vein being about four foot deep and that before he went for Scotland he made ane Essay of Said Copper mine which ore was Sent him up by Said Daniel Peck and it produced out of every ton or ore eight hundred weight of Copper mettle and he believes in Said mine is a mixture of Silver Ore Su Subs
>
> James Hamilton
>
> Jurat 3^{d} July 1716 Coram mo Charles Peers Mayor.

At about the same time, John Haldane of Gleneagles, a brother-in-law of Sir John, but a firm Hanoverian, made the matter known at court. Another brother in law, Patrick Campbell of Monzie, broke the news to Lady Erskine who was definitely not amused, as revealed in her letter to Sir John of the 8th of July:[3]

> ... I had a letter from our freind at London & he tels me he ha's writ to you of the discovery James H ha's made of Mr Nabits affair it ha's greivd me very much & it is no small satifaction that it ha's not faild by any neglect of mine but he certainly designd to comitt the villainy & went away with that veiu for nothing I could doe could make him stay God in his wise providence ha's orderd it & I must submitt but it is a great tryal I have done already what was fit to doe upon such ane exigence & my freind will doe all in his power at London but what will be the end of it God knows but I am not yet alltogether without hop tho I must own my grounds are but small I dare nott writ so plainly to you of it as I incline least it should miscary & doe ane injury on that particular ....

However, the Erskines were quick to turn the situation to their advantage. They exploited an old Scots law dating back to 1592, which said that a tenth of the proceeds of any mine of gold or silver should go to the king.[4] It was argued that pardoning Sir John would enable him to reveal all his secrets about the mine as well as simplifying the legal situation. He could then continue mining with a tenth of the revenue going to the crown. Patrick Campbell explained the plan to Sir John's brother, Robert.[5]

> Dr Sr, – You will no doubt have heard that your brother Sr John was involv'd in the miseries and misfortunes of Scotland last year. He went (as we generally believe here) to ffrance in february last, and an incident has hapnd with regard to him lately, which if well improven may I hope extricate him and his family out of the present misfortunes, the case is this:
>
> One Hamiltoun who came lately from Scotland brought some rich ore with him, and tells of his having found itt at Alva, wher a mine of that sort has been wrought for some time, that ther are some rich veins yett to work, this naturally [caused] an inclination in the ministry to send to try what was in the matter, some time was lost in thinking of the proper persons to be sent, and when these were thought of some difficulties arose about the laws of Scotland with regaird to the mines. The kings advocat gave his opinion, but it was not agreeable to the prepossession that people here were in on that subject, and I'm told some lawyers here sent their opinion on that head from Scotland differing from the Advocat. The use that your Brothers friends here made of these matters was to propose Sr John's remission as a good expedient to remove all difficulties about finding out such mines as were in his grounds, wherof they believ'd he had more knowledge than any body els, and 'twou'd putt an end to all controversies as to the law and right to the mines, for if the right was in the King, ther was no more to be said about it if it or any part of it was in Sir John, then the King was to be umpire, in any difficulty or controversy could arise on that head. My Lord Townshend entr'd into the motion and propos'd it last Thursday in the cabinet wher it was aggreed to. The Prince allow'd Sr John's friends to write to him to leave France, and on Friday V.

> Townshend wrote as I understand it to the King for a warrand or allowance to pass Sir John's remission. In the meantime those who are to be employ'd as Commissioners are it seems exceedingly impatient, and want to be gone to make ye search, and my Lord Townshend is likely to give way to it. I doe apprehend some dangers from this, for if [they] find nothing wher they imagin'd mountains of silver, I'm afraid they'll be the more cool about passing the remission. If they stumble upon any thing that's valuable, then some scruples will very probably be suggested to the ministry why they shou'd not pass the remission, and therby take upon them ye ill will and displeasure of those people who are exceedingly zealous that the estates and effects of the Rebels should goe towards the payment of the publick debts.
>
> My Lord Townshend has allow'd your nephew Sr Hary Stirling to goe over to find Sr John to deliver him the letter wch I am allow'd to write to him, and if in the course of this affair it be necessary to send him to Hanover or any wher els, I think its good you have him with you, for besides the pleasure off seeing you, for which all your friends will envy him, he can tell you more of that affair and of your friends in this country than I can write. 'Tis true My Lord Townshend does not seem to apprehend any difficulty or delay att Hanover, but if you find it otherways, you'l no doubt think of all the proper expedients to remove them and prevail with Sir John to act the part which his friends doe expect of him on such ane occasion, that is to comply with the gentle terms propos'd, and I do not apprehend any other, because they were not mention'd when V. Townshend [wrote] the letter which Sir Hary carries. I expect your Bror Charles and sister Katharine in town by the end of this or the beginning of next week, and how unwilling soever you may be to write, I believe we here shall not scruple to give you frequent trouble till we know the fate of this affair. I am etc.

The government fell for it and decided to send someone to inspect the mine. At first Sir Isaac Newton was approached, 'but he represented himself unacquainted with those matters & declined recommending any body in point of skill'. It was proposed instead to appoint one Dr Justus Brandshagen, 'who had been imployed in working the kings Mines in Germany' to inspect the mine.[6] On 20th August a royal warrant was issued authorising payment of £60 to Brandshagen and £30 to James Hamilton 'to go down to Scotland to work or try the working of this Mine'. Brandshagen was also to be paid 20s per day and Hamilton 10s per day from their arrival in Edinburgh until the completion of their task.[7]

They received their £60 and £30 on the 31st of August, but James Hamilton insisted that his brother Thomas should come too as he did not feel able to carry out the task himself.[8] Accordingly, Thomas was appointed and employed at the same rate as his brother, Sir Isaac paying him an advance of £30 on 5th September.[9] On 25th August Newton sent to the treasury a list of instructions for the inspectors.[10] The instructions were signed by the Prince Regent and issued on 3rd September. The Earl of Lauderdale was also asked to assist the inspectors upon their arrival in Scotland.[11] The instructions were:[12]

By his Royal Highness the Prince Guardian of the Kingdom.

George P.C.R.

Instructions which we (in his Maties name) do will and require to be observed by Justus Brandshagen and his assistant James Hamilton in the Survey and tryall which they are to make of the Mine with the Mountain about it called Sr John Erskines Mine in the Parish of Alva five miles from Sterling East and by North.

You are to repaire with all convenient Speed to the s^{d} Mine, And (having the Assistance of John Haldane of Glenagles Esqr who is willing to Encourage and forward this business) you are to break off or cause to be broken off from each of the two Veins of Ore which are represented to be in the said Mine about Six or Eight pounds of Ore, the breaking off of the Said Ore being to be performed in the presence of the Said M^{r}

Haldane, and any One or more of his Sons together with Mr Drummond Warden of his Ma^{ts} Mint at Edinburgh, or any other person; whom the Earl of Lauderdale General of the said Mint Shall think fit to Send thither And you are to take Care, that the Ore which Shall be broke off in the presence of the Said persons be immediately Sealed up in two papers with the respective Seals of the Said persons each paper to have an Inscription denoting what Vein the parcel of Ore contained therein is taken from, which Inscriptions are to be Signed by your Selves and the aforenamed persons as Witnesses, And then to be safely packed up and Sent by the best and Speediest Conveyance to the Lords Comissioners of his Ma^{ts} Treasury in Order to be Assayed in London.

You are also with the assistance and in the presence of the persons aforenamed to break or cause to be broke off other peices of Ore from each of the Said Veins and make Assays thereof upon or near the place where the Same Shall be taken and repeat the Said Assay Once or twice if need be And the Assays so made with your report thereupon are to be immediately Sealed up (the Assays upon each Vein in distinct papers) with the respective Seals of the said person, And to be inscribed by them in Testimony that they were made before them, which Assays after being so Sealed Witnessed and Secured as aforesaid are also to be sent by the same Conveyance to the said Comissioners of his Majesties Treasury.

You are to give a discription in Writing of the Said two Veins as to their breadth depth and distance from one another and to Report which way they run and what sort of Earth or Stone they are lodged in, and what is the depth of the Mine and the distance of each Vein from the Surface of the Mountain where the Mine is lodged, And whether in that Mine there be any beds of Silver or Copper Ore besides the Said two Veins.

You are in the presence and with the Assistance of the persons aforesaid to Enquire after and Search for Sundry Casks which We are informed were filled with about Forty Tons of Ore dug out of the Said Mine in the time of the late Rebellion and buried in the North west side of the Lady Erskines house and upon finding of the Said Ore to make an Assay and Report thereof in the presence of the persons aforenamed, And then to

Secure the remaining Ore in such place and manner as you with the persons aforenamed or the Major part of you Shall think fit.

You are carefully to View the Burn or Channell made in the South Side of the Mountaine by Floods running down about three or four Furlongs Westward from the Said Mine and See what Sparrs and other Signs of Minerals are therein And in case any Ores of Minerals or Mettals be found there you are in the presence of the persons aforenamed to Assay the Same, and report the produce, Witnessing Sealing up and discribing all Assays to be made by you in like manner as afore directed.

You are Carefully to View all Sir John Erskines part of the Mountain, And Observe what Signs of Minerals appear any where above ground, and Report what you find, As also any Credible Informacons that may be given you concerning any other Mines of Copper or Lead in that Mountain.

You are also to repair to a Copper Mine which lyes (as We are informed) about two Miles Westward from the aforesaid Mine, And procure some pieces of that Ore and make Assays thereof in the presence of the aforesaid persons and See how much Copper and how much Silver it holds.

Lastly you James Hamilton are to be Ayding and Assisting to Justus Brandshagen in all your Observacons Tryalls and experiments, And both of you to use your best Skill and diligence in performing the Same And in case there be any other Mines within three or four Miles of the Mine hereby meant to be tried you may repair thither and Examin what Silver the ore may Contain, And you are to make a return in Writing of all your proceedings herein to the Commissioners of his Mats Treasury with the utmost Expedition that may be.

Given at his Ma^{ts} Court at Hampton Court the Third day of September 1716. In the Third year of his Ma^{ts} Reign.

The brothers left London on the 5th or 6th September and arrived in Edinburgh on the 14th. In the interim, One of Sir John's former servants, John James, appeared in Bristol, sporting a piece of ore and alleging that the mine was 'of many millians value'. His letter is printed in full in chapter 1. He was too late to get involved in the inspection, however, and, as his name hardly appears again in official papers, it seems unlikely he benefited from his treachery.

In the meantime, Dr Brandshagen was having a hard time. He set sail on 10th September, arriving in Scotland on the 13th October; an unpleasant voyage being followed by delays on arrival:[13]

> I got not the money out of the Exchequer till the 31st of August, and my instruction not till the 5th of September, and then I could not get a ship for Scotland to carry over myself till the 10th of September. when I sealed with these things I had brought from St. Catherines, and was three weeks and two days in a dangerous voyage, and in two storms we lost two masts, and were trice driven upon the Sand Bancks. When the ship was repairing all the passengers went on shore, which was very chargeable to me .... The 15th of October I came to Edinburgh and had a conference with the Right Honourable the Earl of Lauderdale, as also with Mr Haldane of Gleneagles, and with Mr. Drummond as commissioners with him. I found in this conference that no provision was made neither for myself nor the two Hamiltons, nor for any incidental charges. My Lord Lauderdale and the other commissioners as well as myself had that time advice from London that Sir John Areskin had obtained his remission upon the fundament that he should show me his mine himself, and it was then agreed amongst us not to go to the mine before his arrival, which was shortly after, and he was then every minute ready to shew me his mine; but for want of money neither myself nor Hamiltons could then go, I wrote to London about it but got no answer.

After borrowing £67, Brandshagen and his party were able to proceed to Alva. A journal signed by the Earl, Haldane and Drummond describes what happened upon their arrival on the 13th of November, the anniversary of the Battle of Sherriffmuir:[14]

> Journall of the proceedings relating to Sir John Erskines Silver Mine

and the Copper Mine att Athrie in pursuance of His Royall Highnes the Prince Guardian of the Kingdom His Instructions dated att Hampton Court the third day of September 1716.

Tewsday Nobr: 13th 1716.

The Right Honourable the Earl of Lauderdale Generall of His Majesties Mint att Edinbrugh, John Haldan of Gleneagles Esqur. Mr William Drummond Warden of His Majesties Mint att Edinbrugh Doctor Justus Brandshagen and James Hamilton being met this day att the House of Alva.

His Royall Highness the Prince His instructions to Justus Brandshagen and his assistant James Hamiltoun for the survey and Tryal which they were to make of the Mine with the Mountain about it called Sir John Erskins Mine in the Parish of Alva were read and considered.

Sir John Erskine having signified that he was willing to shew us where the said Mine was It was agreed that wee should goe together with Sir John first to the place wher the Mine was, And wee went with Sir John accordingly, and he shew us the place, And there it was asked at the said James Hamilton, if that was the verrie place wher the silver Mine mentioned in his affidavits was, And he declared that was the place, and the onlie place that to his knowledge anie Ore had been dug out of. And the shaft or sink of the Mine and Levell which entred to the same from beloe being entirely filled up with earth and stone wee considered immediatly of the propper methods to sett about to have both the Levell and shaft cleared.

Then Sir John Erskine conducted us to the burn or Channell made in the south side of the said Mountain by ffloods running down, about three or four furlongs westward from the said Mine, where Doctor Brandshagen and James Hamilton made a little survey of the Burn or Channell and found some peeces of stone they called sparr which they shew us, But the Burn being of difficult access by the steepness of the Rock over which it falls, the water running in a Gutter of several fathoms depth in the Rock some materialls were first necessarry to be provyded in order to a more particular survey thereof by the Doctor – From thence we returned back to the House of Alva, Sir John Erskine caus'd provyde convenient lodgeing for the Doctor and James Hamilton near by the Mine, and sent his servants to finde out workmen for cleanning the Mine who brought six *viz* Robert Morison William Taylor Andrew Burn John Mows David McNaire and John Drysdale, whom wee engadged at ten pence pr day, being the ordinary wages to these who digged under ground, in the countrie about, they furnishing picks spades and shovells and were to sett to work next morning airlie.

Wedensday Nobr: 14th: 1716.

The workmen were sett to work to the cleanning of the Mine and Levell airly this morning Doctor Brandshagen and James Hamilton and Thomas Hamilton his brother alongst with them. The Earl of Lauderdale and wee went also to the Mine and remaind with the workmen till two aClock in the afternoon, and they made a considerable advance in cleanning of the Mine of earth stones and water this day so that the Levell was got near cleared and the workmen got in the Length of the shaft beloe. But the earth falling down from the sink into the Levell some boards and trees were necessarry to be provyded for supporting the Roof, that the work men might work in toward the Veins, and orders were given, for getting the boards and trees against nixt morning airlie. In the afternoon James Hamilton with two workmen were sent for, from the Mine to the house of Alva where James Hamilton was desired to shew and condescend upon the particular place wher the Casks containing about fourty Tunns of Ore dug out of the said Mine in the time of the Late Rebellion was buried according as he had mentiond in his affidavit, The said James Hamilton accordingly did shew and condescend upon a particular place at the entrie of the House of Alva on the North West side of the said House, where he said He hade seen the said Ore Hid in sundrie Pyps and Casks, and that ther was not above half a foot of earth covering the tops of them, when they were sett to work at two different places in that ground and Dugg at each of the two places, above a yard deep. The earth where they Dugg appeared to be open and Louse, and a few small peeces of Ore were found amongst the Earth. But no Pyps nor Casks were found. But having considered this peece of ground in which the Casks were said to be Hid it appeared to us that it could not have containd the quantity of fforty Tunns. But Sir John himself conducted us into his Gardine and shew us a place where he said he was informed by his servants since he came home anie part of the Ore which remained that was Dug out of the Mine, was buried.

It being at this time Late all the workmen were orderd to attend nixt morning in order to Digg in that place in the Gardine to see to finde that Ore – And when that should be over it was agreed that wee should repair to the Copper Mine which lyes near fyve Miles westward from the Mines at Alva, and is called the Mine att Ethrie (which is the only mine known in all that countrie) In order to perform there what was directed by His Royall Highness Instructions.

Thursday Nobr 15th:

This morning wee with Doctor Brandshagen, James and Thomas Hamiltons and the workmen with Sir John Erskine went into Sir Johns Gardine, where one of Sir Johns servants pointed to the workmen the place where he said the Ore was hid under ground in barrells, so they immediately turned off the earth and wee discovered six small Casks or Barrells filled to the tops what was in them was presently ordred to be taken out, which was done accordingly And upon full examination by the Doctor James and Thomas Hamiltons, the same was almost entirely found to be nothing but stones that probably hade been brought out of the shaft or sink when they were digging for the Ore, and ther was so little of it that contained any Ore att all, that the Doctor and Hamiltouns gave it as their opinion that it was not worth the charge to endeavour to get out any metall appeared to be in them, and that in ther judgement what stones were there were of no value att all. And so wee thought it quite unnecessary to be at any charge in securing these stones so wee left them upon the ground out of which they had been dug save a few small peeces in which ther appeared to be some Little ore which wee desired the Doctor and Hamiltons to keep to make tryall off. And having informed our selves at Sir John Erskins servant and otherwise in the best manner wee could if there was any more of the ore that had been Dugg out of this Mine Hid or buried any where else about the House wee were told that the rest of the Ore hade been from time to time carried away, but they knew not where. Immediately after wee with the Doctor and Hamiltons took horse and went to the Mines att Ethrie, where the Doctor and Hamiltons surveyed these Mines in our presence and went into the severall Mines that hade been Dugg and surveyed severall places about where the ground hade been laid open in search of Veins, and wee found in each of these places appearances of Ore and found severall peeces of Copper of which the Doctor and the Hamiltouns were desired to gather some small quantity in order to trye the finess of the same, and this copper Mine having for a long tract of years and att different times and by sundry undertakers been wrought and many Pitts and shafts appearing wee ordred the Doctor and Hamiltouns to sett down their own opinion of the appearance of this Mine which is hereunto annexed and signed by them.

Wee went from thence to Stirling together with is about two miles from Ethrie when wee discoursed with the Doctor and Hamiltouns upon the other contents of His Royall Highness Instructions, and gave them orders for their direction in there furder proceedure, and that the Doctor with the Hamiltons should return tomorrow morning to the Mine att Alva and continue Closs there till they cleared the Mine and considered every thing they could discover in it, and that then they should make up ther Furnace in order to make the Tryalls as directed by the Instructions. As also that the Doctor should make what progress he could in surveying the Mountain and Burns or Channells, and that how soone matters were thus farr prepared The Doctor should acquaint the Earl of Lauderdale Mr Haldane of Gleneagles and Mr Drummond and wee agreed upon the Doctors advertisement to sett out again for the Mines, and if the Veins and Ore was found to see the same Assayed and tried and likewiyes a

> tryall made of the Copper That wee might be enabled to make a full and satisfieing Report of all this To His Royall Highness
>
> Lauderdale
> Haldane
> Drummond

Over the following weeks the mine was cleared and furnaces constructed for the assays, and another £100 was forwarded to the doctor and the Hamiltons. In the morning of the 2nd of January, the men assembled at Alva and the inspection commenced. Under the supervision of the others, the Hamilton brothers took samples of ore from six separate parts of the vein. The samples were divided, one portion of each being sealed up and sent to London, the other being assayed by the doctor and the Hamiltons over the following days.

Converted to percentages, the results ranged from 0.1 to 5.8% silver, the ore being richest where the vein was widest. In his report, Brandshagen said of the ore:[15]

> I found it of an extraordinary Nature, such as to my Knowledge Few or None Like have ever been seen in Europe; It consists of Sulphur, Arsnic, Copper, Tinn, Iron, some Lead and good Silver, of all these, the Silver only is to be regarded, for the other Minerals and Metals contained in this Ore, are of little Value, and not worth the Charges to separate and keep them.

He also wrote:

> That with regard to Sr John Erskine himself (Who has been a Witnes to all the Transactions with the Commissioners) he has all along been not only particularly civil and kind to me, by procuring me & my Family and the Hamiltons good Lodging and Accomodation near the Mine, in the Minister of Alva's house, but likewise in accomodating us sometimes with necessary Tools, with Work People, and with a house, where we built the

> Furnaces, and made the Assayes, and he contribute in every thing to facilitat and carry on this Business.

A strange feature of the accounts of the proceedings at the mine is the absence of any questions put to Lady Erskine, even though she had been in charge of the mine and the hiding of the ore in barrels during the rebellion.

Afterwards the inspectors returned to Edinburgh and the doctor began to write his report. On 31st January Sir Isaac Newton wrote to them that Thomas Hamilton should stay at Alva to watch over the mine, but the commissioners at Edinburgh decided to send James there instead. They felt that James was 'unskilled in the business he was sent upon' and preferred to send Thomas to London with the report.[8] The report was a long time in the making, however. It was not until the 19th of February that Thomas and the doctor left for London with it (but not before receiving a further £50 to enable them to make the journey). After all, Brandshagen was being paid a pound a day and the Hamiltons half of this amount for however long it took them to complete the job; they seemed to be in no hurry.

The delay did not go unnoticed and was the subject of much criticism by Sir Isaac Newton who, some years later, was to write:[8]

> James Hamilton pretended to know the Mine & yet they loitered till Sr John Erskin came down into Scotland to shew them where it was, wch was about weeks after. They were above five months in Scotland in executing their commission & might have done it in less than half the time an so much that the Earl of Lauderdal Mr Drummond & Mr Haldane declared themselves ashamed of the delays as appears by Mr Drummonds letter to S^{r} Is. Newton of Feb 16. 1715 [*sic*]. And the Commers of ye Treary were impowed by the Princes Warrant to consider this negligence.

After arriving in London, Brandshagen still 'loitered' two months before giving the report to the Treasury on 29th April, and on May 17th he tried to claim pay at the rate of 20s. a day for himself and 10s. a day for the Hamilton brothers for this period in which he appears to have done absolutely nothing.

Needless to say, the Commissioners of the Treasury were not impressed and overruled the claim, declaring that the execution of the Prince's warrant had finished on the 19th of February.

Meanwhile, James Hamilton was still in Scotland watching the mine to see that no ore was taken away. He lingered there until the summer before being instructed to leave and he was paid £20 for 118 days 'wch is after the rate of 3^{s} 4^{d} per diem. And this would have been thought good wages in Scotland for doing so little'.[8] He later tried to claim 10s. per day for 145 days (£72. 10. 0.) and even tried to sue Sir Isaac Newton in Easter 1724 for the difference (£52. 10. 0.),[16] something which cannot have pleased the great philosopher who wrote:[17]

> He came to S^{r} Is. Five or six years ago for more money & S Is. then told him that if he was not satisfied wth ye 20$^{£}$ he must apply to the Comm^{rs} of the Treary but instead of doing so he lets things rest till circumstances are forgotten, & persons are either dead or dispersed, & then sues Sr Isaac upon pretence of a old promise. & of doing this six or seven years after the Reports when original Papers are lost & circumstances are forgotten & Commissions are changed & Persons concerned are either dead or dispersed.

The outcome of this suit does not seem to have been recorded, but it and the earlier delays, plus the claims for cash, do suggest a mercenary attitude on the part of James Hamilton. The sums of money he did receive were very substantial for what was not a difficult task. He was given £30 on 31st August 1716 and 10s. a day from the 14th September until the 19th February (£79), and £20 for watching the mine for another 118 days. There were also payments of £100 and £50 made to Brandshagen and the Hamilton brothers while they were in Scotland, to meet their expenses.

One cannot be sure just how much their expenses were, but Sir Isaac reckoned that, for his journeys to and fro, James Hamilton would only have needed about £10-£15 out of the original £30.[8] Sir John Erskine also commented, 'The charges & expenses were not extraordinary tho the time lost in doing itt was'.[18] In short, Hamilton probably did rather well out of the whole affair. For comparison, in that period in Scotland, someone with an annual income of £80-£100 would have been considered well-off.[19]

A financial motivation for James Hamilton's betrayal of the Erskines therefore seems possible. He may even have had financial difficulties. It was suggested that the reason the £120 advanced to Brandshagen and the Hamiltons had been spent before they even reached Alva was that it had been used to pay off old debts in London.[17]

There may also have been some conspiracy with John James. Mr James' betrayal came too late to have benefited him, but there is evidence of nefarious goings-on involving him and James Hamilton shipping ore from the mine without permission during the rebellion. An undated, unsigned document, possibly by Sir John Erskine, entitled 'Questions desired to be askt of James Hamilton', includes the following questions:[20]

> Did you employ John James to carry from Alva Mine or from the house of Alva some of the Oare of the Mine in Alva, or did He not doe the same with your privily & knowledge?
>
> Did you cause three casks of Oare to be shipt from Alloa in John James's name?
>
> What was the size and weight of the Casks which was shipt in John James's name? Or by him? Or you in his name?
>
> To whom was those casks of Oare consigned, or what became of them and the Oare?

We already know from John James's letter (see chapter 1) that he and James Hamilton were friends and shared a room in London, so a conspiracy between them to betray the Erskines does seem distinctly possible.

Although the government received Brandshagen's report on the 29th of April 1717, bureaucracy, then as now, was slow; it was not until the 6th of December 1718 that Sir John was granted his 'Lycence to Work his Mines'. This licence gave Sir John and his successors the legal right to work mines in his grounds, sending a tenth of the proceeds to the crown:[21]

> ... [his Majesty] perpetually confirms to Sr John Areskine of Alva in the County of Stirling and to his heirs and Assignes whatsoever All and whatsomever Mines of Gold or Silver, Lead, Copper, Tinn and other whatsomever Metalls or Mineralls which is or may be found within all & every the Lands & Heretages belonging to the said Sr John Areskine lying within the Shires of Stirling and Clackmannan With power to him to Seek and Discover Labour and Work the said Metalls and Mineralls and to Sell dispone or sett the Mines therof in Tack or Few to others his Substenants but with consent alwise of Our Treasury and the Barons of Our Court of Excheq^{r} in Scotland ... paying therefore yearly the said Sr John Areskine his heires and Assignes to his Ma^{ts} and his Royall Successors the just Tenth part of all and Haill the Gold Silver Copper Lead Tinn and other minerals which shall be found and gotten yearly within the Said Sr John Areskines Lands and Heritages ....

Soon afterwards Sir John recommenced mining. It may have been in this second phase of working that the bulk of the ore was raised. As will be explained in chapter 4, the bulk of the orebody seems to have still been in place when Brandshagen inspected the mine. Very few records have been traced regarding output after the rebellion. There is a mention of 270 stone and then a further 180 stone weight of ore being raised in 1719. After deducting the charges for refining it, His Majesty's tenth of this output was estimated to be worth just £20.[22]

Such a small amount seems surprising in view of the considerable quantity of rich ore in full view when Brandshagen carried out his inspection. Henry Kalmeter, visiting on the 5th June 1720, reported that although mining was underway, and a smelting house being built, there was no ore to be seen. A 'considerable quantity of ore of several sorts' had, however, been obtained between the time of Sir John's pardon and Kalmeter's visit.[23]

The next record relating to the mine, rather than of production by Sir John's miners, is of an attempt to lease the mine to one of Britain's largest mining companies, the London Lead Company. Their enquiries soon revealed the story of the mine as shown by their minutes for the 24th of July 1722.[24]

> Dr Wright & Mr Creed reported what they hade Learnt about the Silver Mines in Scotland.
>
> Sr John Areskines Silver Mine is towards Sterling, called Alva, Wrought a little by him, but being Forfeited by his joyning With The Pretender; the Government hade it Vieued and Tried by the order of the Prince When Regent, The report to the Lords of the Treasury Wee Saw, Which was very rich in Severall Tryalls made befor the officers of The Mint in Scotland and signed by them; about which Time Sr John Gott his Pardon at the Request of The Czar; whose Phisiton was a near relation of Sr Johns and So is restored to his Estate again; but being Impovrished by these things has Wrought it but Little Since & has been Perswaded by his Friends to Lease it to Some able Undertakers; Severall have attempted to take them but hitherto Without Success; Either Sr John not Likeing the Persons or the Persons his termes or Some other Impedement. however tis thought if Sr John Liked the Undertakers he would alter his first termes, and would Lett the whole Liberty; which is about 2 Miles long and above one Mile broad, in which Severall Tryalls have been made and a pretty deal of rich ore got; but it Seems not a Great Deal Wrought the soil being pretty hard & Some water afecting Some Part of the Worke; by Sr Johns Lett^rs^ there are Severall veins of Different breadths with Smaller & Large Ribs of ore which are Expected to Grow Stronger in the Depths; The veins being full of Spar and Mine Soil.

It was unanimously agreed to draw up a draft lease immediately. Such arrangements were common in mining enterprises. They freed the owner from the risk and burden of investing heavily in a speculative venture – the leaseholders took the risk – and, in return, they were assured of a share of any ore raised. In fact Sir John himself was already involved in such a scheme at Leadhills where a company, of which he was a partner, worked lead mines with an eighth of the proceeds going to the Earl of Hopetoun.[25]

Such a deal would be understandable if all the visible ore had gone and the only way to find more was to take a chance and invest in digging deeper. Such an undertaking could easily consume any profits made previously, something Sir John may not have wanted to risk. This, in turn, begs the question, what happened to all the rich ore observed to be still in place by Brandshagen? Was it still there, or had Sir John mined it and not declared it? If still there, why lease it out, and relinquish the lion's share when he could mine it himself? After his brush with the law, the awareness of the authorities and, doubtless, plenty of people willing to betray him (as James Hamilton and John James had done), mining and not declaring would seem risky to say the least.

Whatever the case, the gentlemen of the London Lead Company were in no doubt they were on to a good deal. They had samples of the ore assayed and valued at £1000 per ton.[26] Their seal was placed on the agreement on 7 May 1723[27] and their copy delivered to a Mr James Campbell the following month who was to take it to Edinburgh to exchange it for Sir John's copy.[28] Shortly after it was decided to send two of the company's most respected shareholders, Dr. Edward Wright and Mr James Creed, to view the mines in Derbyshire, Scotland and Wales.[29] Their account of their visit to Scotland is a fascinating insight into how business trips were conducted in the early 18th century.[30] After various stops in England, to look at mines and dine with a mayor, their coach arrived in Edinburgh on Tuesday 23rd July.

> Where we were Wellcomed by Mr Campbell of London Sr John Areskine and his brother Mr Char: Areskine Mr Campbell of Monzie and severall other Gentlemen friends of Sr John as also by Robert Wightman Mr Geo; Irving & Mr Drummond one of their Partners & a Commissioner of the Custome house, after wee dined with Sr John Areskine who Entertained us then and the next day very handsomly – and delivered to us the Lease Executed by him to the Company.
>
> Thursday ye 25th wee dined with Mr Wightman & Supped wth Mr Campbell of Monzie Friday the 26th wee dined with Mr Provost Campbell one of their Members for Parliament, a very Civill Gentleman & Friend to Sr John, & in the afternoon wth S^{r} Jn Areskine.

> Wee Sett out for Hoptoun house by invitation from y[e] Earl where wee Lay and were Entertain[d] very Kindly and in an Extreame handsome Manner. The next morning wee had a pretty deal of discourse with his Lordship concerning his Mines and Liberties, and he is very willing y[e] Company should be adventurers with him, on any reasonable terms and ordered his Steward Geo: Sherriffe should attend us at Lead Hills, that wee might view his works & Liberties wch wee did Accordingly
>
> Saturday the 27th wee parted from y[e] L[d] Hoptoun and went with Sr John Areskine to his own house, being in our way at Sterling Complimented wth Burgess Ticketts.
>
> John Newton & Abr[a]: Watson who mett us at Edinburgh attended us to the Lord Hoptouns, and to Sr Johns, Sunday the 28th wee had a great deal of discourse with Sr John the whole day being spent in his house, on Monday the 29th & 30th Wee went to view the Silver Mines.

It is clear from the detailed description that followed that someone, whether the London Lead Company in advance of the exchange of documents, or Sir John's own men, had been very busy. Several new veins had been opened up and levels of up to 20 fathoms length driven along them, as well as various shafts and cross-cuts. These included 'Donnalls vein' 20 fathoms west of the silver vein on which a shaft had been sunk seven yards deep, 'Gold vein' further west still, and 'Smithy vein' 30 fathoms east of the silver vein and running in a northerly direction.

On completion of their inspection, Dr Wright and Mr Creed

> ... were Entertained by Sr John very handsomely and in a very friendly Manner, he is an Ingenious gentleman & Soe well Satisfied and pleased with the Companys proceedings with respect to his Mines, that there is no Doubt of making a fresh agreement with him for his Mines, if the present Subsisting bargain should not be proceeded on at Expiration of time of Tryall, and believe he will come up to London himself, if there should be occasion to make a fresh agreement, he in a Manner promised as much, & Looks upon himself obliged to doe, as the Company have performed

> Soe handsomely, all that was Expected on their part. The ore is Prodigious rich as the Company has been before Satisfied in. Wee had Severall Samples, and some took out of the Mines which being very Weighty and Cumersome to carry round our Journey wee Left to come up by Sea.

Although they left Alva on Wednesday 31st July, the wining and dining continued until the following Monday. During this time the two visitors were made Burgesses of Stirling, at Sir John's instigation, supped with Magistrates of the City of Edinburgh, were 'presented with the Freedom & admission of Guild Brethren, with a great deal of forme and Ceremony', visited the city's water supply, and further supped and dined with assorted bigwigs including Sir John (again) and the Duke of Queensberry – and all at their hosts' expense.

It seems the original agreement was not to Sir John's liking and he approached the Company in March 1724 seeking to amend it.[31]

> Dr Wright Delivered a proposall from Mr Char: Areskine relateing to Sr Johns Mines Vizt:

> That in Stead of £5000:– offered to ye Comp^a^ to be paid Sr John in Lieu of the fine, in case the profitt ye Comp^a^. should make of those mines amounted to ye Sume in any one year During ye Lease; it should be agreed thus Vizt when the Mines comes to yield profitt at suppose £1000: in one year Sr John may receive it as the highest year had yet happened, but when a year of greater profit happens, as Suppose £2000, - or more then Sr John to have the first Sume made up to the Last, and if no greater Sume happens to be got in any one year, then Sr John to have nor challenge and more dureing the lease but when ever any greater profitt happens in any one year; then the Severall Sums received the Comp^a^. to make up the Last Sum to y^t^ as farr as £5000 and noe further.
>
> The Court haveing deliberated upon it, came to the following resolutions y^t^ to shew the great Respect they have for ye proposer & disire to accomodate with Sr John they will Comply therew^th^ provided noe further time be Lost of proceeding on ye Worke, wch is both Sr John^s^ interest &

> our desire, And as a further proof of their Sincere desire to proceed on the further Tryall of those mines and to remove Sr John^{s} Scruples about the makeing up the acco^{t} of profitts for any One year we will give Sr John in Lieu of a fine, 1/4 instead of 1/5 as farm which we apprehend to be more than an Equivalent if the mines ever prove Valuable: and these are our finall resolutions we Desire Sr John Areskines answer to them wth out delay & upon his non acceptance we shall immediately be ready to Exchange deed^{s}, Part friends and wish Sr John bett^{r} tennants.

Gobbledegook, it seems, is nothing new. If this new agreement about how the profits were to be shared seems confusing, then the final agreement is almost incomprehensible.[32]

> That as soon as the Company have gotten ore enough to incourage them to Smelt and refine, Sr Johns Wastes, Slaggs toolls & buildings (wch the Company shall judge Usefull for Smelting & refining) shall be appraised & Valued, and ye Company shall pay Sr John for their share there of, vizt 11/20th parts and S^{r} Jno as partner in Smelting to have 7/20th parts, wch proportion are respectively to be born by the Said Partyrs, of carriage Smelting and refining, & of all Charges attending the Same; this method may prevent any dispute about the wastes and Slaggs, but if Sr John rather chooses to Worke up the Said Wastes, and Slaggs, than accept the appraised price, then he to have Liberty to doe the Same with all convenient Speed and before the Company begin to Smelt and refine, and the Company in that case to pay their share only of the appraised price of the tools, & buildings judged necessary for their use.
>
> Memorandum April 29th 1724. In London it was agreed between Charles Areskine James Sinclair & Pa. Campbell Es^{qr}. on the behalf of S^{r}. John Areskine of Alva Barn^{tt} and Edward Wright on behalf of the Gov. & Comp^{a}. for Smelting down Lead &^{c}. That in Stead of the fine of £5000 to be p^{d} by S^{d} Comp^{a}. to Sr John Areskine as a fine as mentioned in a Certain indenture of Lease dated the 21st December 1722 the Comp^{a}: should pay and S^{r}. John should receive one fourth part of all the Ore therein Demised

as a farm in Lieu of a fine and that Sr John should goe 2/13 parts of the adventure of Mineing in profitt & Loss, instead of the 1/7th mentioned in ye Said Lease and shall bear a proportionable share of carrage Smelting & refineing, as those alterations from the Lease make and Create that is 2/13^{ths} & ye Company 11/13 in Mineing and in Smelting Sr John 7/20 & the Comp^{a}. 11/20 and in case of Sr John^{s} Smelting ye Crowns 1/10, then the Expence to be borne as follows, Sr Jno Areskine 9/20^{ths} and ye Comp^{a}: 11/20^{ths} This new Lease to commence from the 15th May now comeing for 21 years & all other covenants and things contained in the aforesaid Lease to be included in the New, wch are not inconsistant wth these new Agreements & to be engrossed & Executed with all convenient Speed, and in the mean time the Comp^{a}. to prepare in order to Set the men again to worke, at Alva, as soon as they can. This new agreement puts the parts in ye following order

Vizt 2/20 p^{ts} to the Crown

5/20th to S^{r} Jno. for farm or duty, which reduces ye rest into 13^{ths} which are to beare the whole Charge of Mineing Vizt 2/13^{ths} Sr John & 11/13^{ths} ye Company

Intriguingly, this document refers to a lease of 21st December 1722 and mentions setting 'the men again to worke', which would imply that the new shafts and levels observed by Dr Wright and Mr Creed on their visit in September were at least partly the company's work. The 'Wastes, and Slaggs' from Sir John's own working, on the other hand, must have been present in some quantity for the company to express an interest in them. This would suggest some significant earlier activity on Sir John's part also.

How the London Lead Company's involvement ended is unclear, as the company's surviving records cease to refer to Alva after this point. In December 1728 a new 21-year lease, to a 'Company for working Mines & c in Scotland', was drawn up.[33] This may just have been a revised lease to the London Lead Company, although it is not clear. In it Sir John sought one-fifth of the ore and one-tenth of the profit. He also mentioned that he possessed at Alva 'a smelting & refining house with Furnaces & Stamping Mill'.

There is a 25-year gap in the records at this point. However the London Lead Company fared, it is clear that mining was pursued with vigour. When Brandshagen inspected the mine there was but one small working. By the time Sir John, and later the leaseholding company, gave up, the original mine had been considerably extended and there were numerous new shafts and levels on other veins in the glen. Some of these also yielded silver ore (although not, it seems, as much as the original deposit). One of these new levels, driven westwards on the 'Self-Open' vein upstream from the silver mine, was 50 fathoms long[34] (the length of a soccer pitch).

With his fortune from the mines, Sir John embarked on a programme of ambitious improvements to his estate. He made agricultural improvements and enclosures, and had the hillside behind his house planted with both native and exotic trees. He also, at great expense, had a canal dug to facilitate the transport of coal from pits along the River Devon to the Forth. When a neighbour, possibly George Abercromby of Tullibody, remarked of these improvements, 'Sir John, all this is very fine and very practicable, but it would require a princely fortune', Sir John replied, 'George, when I first formed my scheme of policy for this place, I was drawing such sums out of the mine that I could not help looking upon the Elector of Hanover as a small man'.[35]

NOTES TO CHAPTER

1 NA, MINT 19/3/268, Newton to Lord Townshend (undated).
2 NLS, Paul MSS, 5160, f. 5, 3rd July 1716.
3 NAS, GD1/44/7/18, 8th July 1716.
4 Cochran-Patrick, R. W: *Early Records Relating to Mining in Scotland* (David Douglas, Edinburgh, 1878).
5 Paul, R. (ed.): 'Letters and documents relating to Robert Erskine, physician to Peter the Great, Czar of Russia, 1677-1720', in *Publications of the Scottish History Society, XLIV, Miscellany* (1904) vol. 2, pp. 371-430; Letter XIV, pp. 414-17 (undated) (see also Letter XIII, p. 414).
6 NA, MINT 19/3/250, Newton to Treasury, written on a receipt for coach hire dated 22nd July 1721.
7 NA, MINT 19/3/239, 20th August 1716.

8 Ibid., 19/3/248 and 19/3/250.
9 Ibid., 19/3/264, 5th September 1716.
10 Ibid., 19/3/231, 25th August 1716.
11 NA, T17/3/528, 4th September 1716; see also NAS, WRH, Lauderdale papers, RH4/69/26/2, 4th October 1716.
12 NA, MINT 19/3/229, 3rd September 1716. (Another copy in T17/3/530).
13 Ibid., 19/3/226, 17th April 1717.
14 NA, T1/201/96-99, 26th November 1716. (Another copy in NAS, WRH, Lauderdale papers, RH4/69/26/3).
15 NA, T64/235, 29th April 1717. (Other copies in NA, 19/3/221 and NAS, WRH, Lauderdale papers, RH4/69/26/6, these latter also contain a note on the assay method used – RH4/69/26/7: The ore was melted with lead which extracted the silver. The lead was then removed by roasting in a blast of air, which oxidized it leaving behind pure silver).
16 NA, MINT 19/3/240, Easter 1724 (in Latin).
17 Ibid., 19/3/?249 (undated) (adjoins documents 248 & 250).
18 NLS, Paul MSS, 5160, f. 10 (undated).
19 Paul, R.: 'Alva House Two Hundred Years Ago: part II', in *The Hillfoots Record* (10th April 1901, p. 3.
20 NLS, Paul MSS, 5160, ff. 6-7 (undated).
21 NA, T17/5/8-9, 6th December 1718.
22 NAS, Exchequer Records, E 307/2, nos 108-9, 14th January 1719/20.
23 Smout, T. C.: 'Journal of Henry Kalmeter's travels in Scotland, 1719-1720', in P. L. Payne (ed.): *Scottish Industrial History: a miscellany* (Scottish History Society, 1978), 4th series, 14, pp. 1-52.
24 NRS, LLC Court Minutes, 5, 24 July 1722, 188-190.
25 NLS, Murray of Stanhope MSS, ADV MS 29.1.1.I, ff. 119-121; copy of a letter to Mr Campbell of Calder by Alexander Murray dated Edinburgh 27th August 1722.
26 NRS, LLC Court Minutes, 5, 18th December 1722, 214.
27 Ibid., 7th May 1723, 250.
28 Ibid., 11th June 1723, 255.
29 Ibid., 25th June 1723, 257-58.
30 Ibid., 17th September 1723, 279-84.
31 Ibid., 31st March 1724, 332-33.
32 Ibid., 28th April 1724, 338-40.
33 NLS, E-M MSS, 5098, ff. 6-9,12th January 1729.
34 NLS, E-M MSS, 5098, ff. 26-7, 1753.
35 Allardyce, A. (ed.): 'Scotland and Scotsmen in the Eighteenth Century' (Wm. Blackwood & Sons: Edinburgh & London, 1888), vol. II, p. 111.

GD1/44/7/13

My Dearest Life

I receiv'd Yours of the 20th & another of the 29 of Nov which wer both most acceptable, butt they had both been long by the way for itt was the 5 of De befor I receiv'd the first, You are much mistaken in thinking I was displeasd with you for leaveing this con[illegible] I doe assure you I thought itt a lucky providence & tho I was in p[illegible] for not hearing from you yett itt was easy to bear in comparison of what terror I must have had if you had been in the danger some others of our freinds have been in, I suppose you know all our difficultys from better hands long Eire now & by that you may guess the torment & fear & terrible Horror I must be in for you & many others if I had known your adress I had writ to you three weeks agoe & begd of you to stay where you was till you saw how things wold be I writ to your Brother in hops he wold learn itt from some att Edr butt he told mee he could nott & you was soon exepected & I was so far from wishing you soon back I was afraid to hear of your return, I pray God send a happy end to all for I am just where I was & my hops are still very faint that person you mention in yours not being come yett your children are very well & all your other freinds I doe nott wish to hear you are returnd butt when you doe pray God you may be saffe which is the earnest wish off her who is intirly

Yours

De 10

I am better then could be exepected all things considerd if you can have any reasonable pretence to stay doe not come by any means Mr Peck gives you his most humble service as does Aunt B & I

Letter from Lady Erskine to Sir John,
dated 20 December 1715, expressing concern for his safety
and suggesting it would be dangerous for him to return.
NATIONAL ARCHIVES OF SCOTLAND (GD1/44/7/13)

Charles Erskine, Lord Tinwald, Lord Justice-Clerk.
Portrait by Allan Ramsay, 1750 [detail].
SCOTTISH NATIONAL PORTRAIT GALLERY

CHAPTER 3

A SECOND BONANZA: THE DISCOVERY OF COBALT

SIR John died in 1739 of injuries sustained in a fall from his horse. The estate passed to his son Sir Henry or Harry Erskine, baronet and member of parliament, and later lieutenant-general in the army.[1] (It is possible the estate may have passed first to Sir John's eldest son Charles, but he died in 1747.) In about 1749, Harry sold the estate to his father's brother, another Charles.

Charles Erskine, Lord Tinwald (1680-1763) was the most able of Sir John's brothers. Pursuing a career in the legal profession he rose to the position of Lord Justice-Clerk in 1748. He was married to Miss Grierson of Barjarg by whom he had many children, one of whom, James (Lord Alva and Lord Barjarg) enters the story later on. In order to buy the Alva estate, he had to sell his estate at Tinwald, near Dumfries. He invested much expense in putting the house and estate in order, but was able to spend only a few weeks a year there while on holiday.

Charles was a well-liked man, described by Ramsay of Ochtertyre as 'possessed of excellent talents, which were improved by culture, and set off to great advantage by graceful persuasive eloquence in a strain peculiarly his own'. He was witty, polite, even-tempered, and his oratorical skills earned him the nickname 'Sweet-lips'. He did, however, have a strong dislike of Jacobites and never missed an opportunity to have a go at them.[2]

Whilst not sharing Sir John's enthusiasm for the Stuart pretenders, he did share some of his speculative spirit. He suggested, 'If one could turn over the Ochils like a bee-hive, something might be got worth while'.[2] So it was that in about 1757 he decided to rework the silver mine, which had lain idle for many years.

A John Williamson, from Leadhills, was appointed overseer of the mines and work commenced in March 1758. Having been neglected for so long the workings were in a poor state, as Mr Williamson described in a journal he kept. For the period Wednesday 29th March to Monday 3rd April, he wrote: 'Employed in drawing out Rubish from the old workings and discovered the Mouth of a Drift which goes Northward, but could not have access into it for making Observations as it was full of Water to the Roof.' They then tried to the west and 'in two hours working discovered another Drift, which goes to the Westward, likewise full of Water to the roof'.[3]

They spent Tuesday 4th April clearing the workings and searching for the mouth of the general level, but they did not find it until the following day. Needless to say, this too was blocked with debris and completely flooded.

In due course the workings were cleared and mining began. Old workings were extended and new ones commenced on promising veins. It was resolved to drive a level for 14 fathoms along 'Lead vein' and by 15th January 1759 there was only one fathom to go.[4] They now wanted to sink a shaft or sump on Lead vein, but this would require 'a turn Head, Eighte foot long and seven foot wide',[4] presumably to accommodate the hand winch that would be needed to haul rock up from the sump. A month later it was finished, but the miners wanted £3 for their efforts, rather than the 40 shillings John Williamson was offering them.[5] By 18th June 1759 'the sump was finished upon Saturday last, which makes in all three fathoms sunk upon lead Vein'.[6] Mr Williamson kept detailed accounts for this particular piece of work which provides an insight into the everyday costs of mining:[7]

Acco^{n} of the Charge of sinking at the Lead Vein.

To Rooming the Turn Head at £1..10 p^{r} fa[illegible]			
	7	-	-
To sinking one fathom @ £1..4	1	4	-
To sinking the 2 fathom	2	15	-
To sinking the 3 fathom	5	10	-
	16	9	-

To 30 lb of powder @ ½ p^r lb £1..15				
To one stone of candle	8	2	3	-
		£14	6	-
To three men at the pumps £2..3..				
To 2lb of candle @ 6 p^r lb	6	2	9	-
		£16	15	-

N.B. The charge of Makeing and keeping the pumps in Repair with leather is not come to hand yet, and therefor could not be Marked in the above Acco^n [illegible] Excepted

attested Jn Williamson

One of the new works was another shaft several fathoms deep, this one sunk directly below the site of the original bonanza and a level driven from the bottom of this shaft. To drain these workings, a level was started lower down the glen. The miners had advanced only a short distance when they encountered a large mass of ore described as having the colour of peach blossom. In the hope that it may be silver ore, samples were dispatched to the eminent Scottish chemist Joseph Black in Glasgow for assay. His reply dated 17th January 1759 was the first identification of cobalt at Alva:[9]

Sir

I am sorry It has not been in my power before now to give you an account of the mineral you sent me last I was partly hindered by want of leisure & partly by an accident which befell my first trial & obliged me to begin it anew so far as I can discover it contains but a small quantity of Silver, about an ounce averdupoise in the 100 lb w^t. But it is not on that account unworthy of attention It is a true Cobalt or fossil of that kind which affords the Saffer or blue for Porcelaine & c. & likewise the Smalt blue & which is sold at the Saxon mines, as dug up, at the rate of from 14 to 30 livres the

100 w^{t}. according to its quality – Yours so far as I can judge seems very good but if you find any considerable vein of it from which it may be dug up by hundreds I shall make some assays of it by making it into Smalt & Saffer upon the last of which I shall procure some trials to be made by painting some delft ware & comparing the colour with ye Saffer used here – It is observed by the miners that the Silvery cobalts even of the same vein are very different in richness & where they are wrought for Silver it is usual to assay every particular heap of ore before it be sold.

If therefore you meet with any pieces that have a particular look you may send them to me & I shall try them - & at any rate I shall take it as a favour if you will be so good as to lay by for me any varietys in the Specimens of it that may occur because it is a fossil of the rarer kinds – The peice sent me contained a little copper but not a particll of Iron that I could observe I am Sir yr. Most humble Servt

Glasgow 17th Jany 1759 Joseph Black

The smalt and saffer (also spelt 'zaffer' and 'zaffre') he refers to are preparations made from cobalt; they were used to impart a deep blue colour to glass and porcelain, this being the main use for cobalt in the 18th century. Zaffer was made by roasting cobalt ore to form cobalt oxide. Fusing the zaffer with sand and potash turned it into an intense blue glass, which could be ground to a powder. This was smalt and was the preferred blue pigment for the manufacture of high quality blue porcelain. Nowadays cobalt also finds uses in magnets and special alloys and is known to be a common constituent of silver deposits of the type mined at Alva. Indeed the Saxony cobalt Joseph Black referred to was a by-product of silver mines.

With the mines now known to contain a valuable commodity, they were swiftly put under lock and key.[9] Meanwhile word spread about the discovery of cobalt. On 19th April one Andrew Crosbie of London wrote to the Erskines, 'I wish you joy of your vein of Cobalt Ore for such it appears to be from the Experiments made here as well as from those made in Scotland'. He said that over £200,000 was spent annually on importing cobalt.[10] With no other indigenous supplies of the commodity, the prospects looked good for the Erskines.

Things seemed to have moved slowly at first. The earliest request in the Erskine family papers from a porcelain manufacturer for samples of Alva cobalt is a letter from Will Davies of Worcester, dated 6 December 1760.[11] Although samples were despatched to him, it seems the Erskines were exercising caution. In a letter to Mr Davies dated 26 August 1761, it was explained that 'our principle Object at present is to trace the vein that carries this Minerall and to secure the prospect of a considerable quantity of the Oar, before we launch out with great Expense in preparing for the manufacturing it'.[12]

In August 1760 the waste from the old silver mine was also reported to contain cobalt,[13] although when it was first observed is not recorded. With such encouraging indications Charles Erskine, together with some relatives and friends, set up the Cobalt Mining Company in 1761. A list of 13 'Articles of Agreement' was drawn up as follows:[14]

> Copy Articles of Agreement of the Copartnery for Raising & Manufacturing Cobalt and other Metals and Minerals on that part of the Grounds of Alva in Stirlingshire Called the Silver or Dovecot burn belonging to Charles Areskine of Alva Lord Justice Clerk
>
> 1mo. The Said Copartnery is to Consist of Ten Shares of £300 St^{r}. each, to be paid in Manner aftermentioned
>
> 2do. If any partner herewith Subscribing Shall Refuse or delay to pay in their proportions ffor the Space of Six Months after being certifyed of a Call being Ordered by the Company, Such partner Shall fforfeit his Shares to the other Copartners.
>
> 3^{tio}. a Lease to be taken for the Term of 30 years from the 1st Day of January 1761 of that part of the Grounds of Alva which Lye upon the Said Silver or Dovecot burn from the Old Smelthouse on the South, to the Source of the Said Burn on the North, And betwixt two Lines paralel to the Course of the Said Burn on the East & West at the distance of 200 fathoms from

it, ffor Working Mines, Digging for and raising all Metalls and Mineralls. Reserving to the proprietor the Use of the Surface and Damages for the prejudices done thereto by the Works, to be Ascertained by persons Mutually Chosen. With privilege to the Leasees in Case they Shall drive on a Vein to Any part of the Boundarys above discribed, which at that place does yeild Metal Sufficient to defray the Expense of Working and to pay the tack dutie, to follow the same While it Continues to bear Metal to that quantity But no further providing always that in no Event the Work Shall be Carried to the Eastward of the Old plantation of ffirrs at the head of the West Avenue.

4^{lo}. The Tack duety to the proprietor to be One ffourth of Silver Ore during the Whole Lease, One Eight of every other Ore or Mineral for the ffirst Seven years, And for the Remainder of the Lease, One Sixth of every ore, Excepting Cobalt which during the last Sixteen Years is to pay One ffifth. All Ore to be Esteemed Silver Ore when it is found upon trial to Contain a Twentieth part of Silver; And all Ore to be Esteemed Cobalt, When the Cobalt is double the Value of every other Metal contained in the Ore.

5^{lo}. That the Dutie shall be paid in Crude Ore. But in Case the Company Shall Smelt or Manufacture themselves, They shall be Obliged to Manufacture the Tack duty as they do their own Ore upon being paid the Expenses of So doing.

6^{lo}. The Company to begin Working, by Carrying up a Level from the ffoot of the Lowest Waterfal, to the place where the Cobalt has been discovered (being about 70 fathoms) by three Shifts of Men at least, if air permit, And to Carry it on as dead as possible, without any unnecissary Delay, Or make other Trials in other parts of the Said Grounds.

7mo. That the Company shall have Equal to Eight pickmen Imployed for at least Six Months in the year And if they shall not Work at the rate of 1456. Days Work of One Man for each year the Lease shall become Void.

8vo. The General plan of the Works and other Business of the Company to be fixed, and Directed by ffour quarterly Meetings In all which every Sharer Shall have a Vote and shall preside by Turns, The president to have a Casting Vote in Case of a Division: And Sharers necissarly Absent May Send Proxys to Any of the Copartners.

9no. That the first General Meeting Shall Appoint a Manager for Inspecting and directing the Works, Keeping the Companys Books, Issuing Money to the Overseer And otherways Managing the Companys Affairs Who Shall lay before Every Quarterly Meeting a State of the Works and their Accounts.

10mo. That the Manager shall lay before each Quarterly Meeting a plan of the Works proposed for the three following Months And no New Works Shall be Undertaken but by the Authority of the Said Meeting.

11mo. That at the first General Meeting Each partner Shall pay Ten [per]-Cent of his Capital to the Manager, Who Shall be then Chosen, to be Issued by him as the Companys affairs Shall Require.

12mo. That Three partners Shall be a Quorum at the Quarterly Meetings for the Ordinary Management of the Companys affairs: And that no General Rules or By Laws Shall be binding But what Shall be Agreed to by at least Seven of the partners or their proxies.

13mo. That the partners Shall not be at Liberty to Assign their respective Shares in this Copartnery but with Consent of the Company.

We Subscribers Agree to the foregoing Articles And Bind and Oblige Us Our heirs & Successors to perform and Implement the same to each other and to Execute a formal Contract in the Terms before Express And Consent to the Registration hereof in the Books of Council & Session Or other Court Books Competent That a Decree of Registration be Interponed thereto that Letters of Horning on Six Days Charge And all other Execution necissary be directed thereupon in form our Effiers and to that Effect we Constitute

Our proxes In Witness whereof &c. The Articles are dated the 11. 12. & 16. of March 27. of April & 31 December 1761. And 2. Febry 1762. – And the Subscribers are –

1. Ch. Areskine of Alva Lord Justice Clerk
2. Ja: Erskine of Barjarg Esqre. One of the Barons of Exchequer
3. Alex^{r}. Sherriff merch^{t} in Leith for himself & James Guthrie Merch^{t} in Ed^{r} in Cop.
4. Geo: Muir Writer to his Majestys Signel
5. John Fordyce Merch^{t} in Edinr ffor Ffordyce Grant & C of London
6. John Stephenson Merch^{t} in Hull
7. Sir Harry Erskine Baronet
8. Nicholas Crisp Esqr. Of Bow Church yard London
9. Andrew Crosbie Advocate
10. John Campbell Casheer of the Royal Bank Ed^{n}.

The eighth subscriber on the list, Nicholas Crisp, initially proved to be a great asset to the company. His skills as both a jeweller and a potter enabled him to assay ores for precious metals and also experiment with the cobalt at his own pottery at Vauxhall (then in the Parish of Lambeth, London).[15] He was also an expert on cobalt, having written an important treatise on the subject for the Society of Arts in 1758. A copy of this is preserved in the Erskine family papers and describes in detail the furnaces and methods used in the processing of cobalt.[16] Not surprisingly, many samples were sent to him for tests, and 27 of his letters to the Erskines still survive providing an important archive for those interested in the story of the Alva mines.

The first of Crisp's letters, dated 17th September 1761, suggested checking the old silver mine waste for cobalt (although this had already been done). He also reported that the Alva cobalt was as good as that from Saxony and that the prices paid for zaffer in London were from 10 to 20 shillings per pound.[17]

He wrote again on 26th September with the results of tests on ten ore samples from different parts of the mines. These were very variable. Some gave good blue colours, others just a faint blue, some only gave brown. Mr Crisp was concerned that impurities in the ore were spoiling the colour and urged that the best quality ore be carefully separated from the rest. He finished with the comment, 'another property I observe in these Ores, and in the Zaffre made from them is that they are apt to spread in the Fire, not give a clean distinct line, well defined, but an uneven, or as we term it a Wooly line'.[18]

On 13th October 1761 Nicholas Crisp wrote again, apologising for the lack of progress with his experiments as he had 'been confined by the Gout, and since by a disagreable Cold'.[19] He said that the ore contained 'Metallick particles' and that it was in these that the colour resided. The 'peach:blossom:colour' also present was the result of the action of the air on the metallic particles. He had also arranged for a Cornishman to travel to Alva and inspect the veins.

His letter on 1st November 1761 carried more exciting news. He had conducted further tests on some of the ten samples mentioned before and found that one, from the old silver mine waste, contained about '1/70 Silver or about 28 – lb in the Tonn W^{t}:' while that from the new cobalt vein contained, after roasting, 29 pounds of silver per ton of ore. 'It is a great consequence to us to think how to get the Silver out without destroying the Cobalt. for this will make a vast difference in the value of the Ore,' he wrote. He finished on a note of caution: 'I think also the affair of the Silver contained in the Ore should not be publickly known unless it is so already.'[20]

Just four days later he wrote again with the results of tests on the remainder of the ten samples. Every one of them contained silver, some as much as 22 lbs per ton.[21] As even a few ounces of silver per ton could often be recovered economically this was an important discovery, but there was more to come. On 23rd November 1761 Crisp wrote, 'I find also in the silver I have extracted from the Ores nearly 1/12 of Gold. but on this Quantity I do not depend. but there certainly is Gold in some considerable quantity'.[22]

The presence of precious metals did not distract from the value of the cobalt, however. Mr Crisp continued his work on the latter and estimated it to be worth £100 per ton.[23] He wrote at length about how best to treat the ore and described the facilities for doing so at his Vauxhall works. He was also pleased that it had been decided that the old silver mine waste should be processed for cobalt, but felt that on its own this would not be 'sufficient to begin a manufacture'.[24]

Meanwhile the miners were busily sinking shafts and driving levels in their searches for more cobalt. In a letter dated 28th September 1762, John Williamson describes progress on the silver vein, a shaft 12 fathoms deep, and states that four men were working on the cobalt vein.[25] On 25th October 1762, he wrote again to say that work was continuing with two men following a string (a narrow vein) 'to the east from the silver vein'. This consisted 'of a small seam of ridish clay in widness two inches, to which there joins at present, a rib of rider four inches broad'. He also referred to 'the high drift on the silver vein ... points eight degrees north east ...'. The work was not without problems, however: 'The shaft has gone very slowly on for sometime past being very cross & hard, and since the weather sett in wet it is with difficulty that she cane be wrought with the watter coming down upon them from the day.'[26] One thing that is consistently absent from his progress reports is any discovery of new deposits of ore in spite of all the miners' labours.

During this period occasional interest in Alva cobalt was expressed from the pottery industry. A William Duesbury requested samples in a letter of 17th February 1762,[27] and a year later samples were also sent to Liverpool.[28] James Erskine took over responsibility for the mines after his father died in 1763, while Crisp continued his experiments in London uninterrupted.

Crisp's letters were frequent and optimistic. 'The Cobalt vein,' he declared, 'contains very good lead, very good Copper and three different kinds or such of very good Cobalt.'[29] In March 1763 he said he was ready to process half a ton of cobalt ore a week and asked for a ton of ore.[30] A week later Crisp reported finding gold again in his assays, but he was having problems with his crucibles cracking.[31] By May 1763 he was making zaffer as good as that from Saxony.[32]

Considerable quantities were now being shipped to Mr Crisp. There survives a 'Note of Casks sent to Mr Crisp at London from Alva on board Lovely Anne at Leith John Smith master', dated September 1763.[33] This consignment consisted of 18 casks of both 'Cobalt oar' and smalt, six of them from the 'old workes'. That some of these casks were of smalt shows that, by now, the company was doing some manufacturing of its own. How much each cask contained is not recorded, but other references to casks of ore suggest that they typically contained one or two hundred weight.[34, 35]

Sadly for Nicholas Crisp, has optimism was about to be cruelly cut short. His considerable technical skills were not equalled by his business ones and his world was about to collapse. A combination of poor investments, mismanagement and living beyond his means, resulted in him being declared bankrupt on 17th November 1763. He had debts in excess of £18,000.[15]

In spite of his personal difficulties, Crisp struggled on. He wrote to James Erskine on 20th December 1763 with the results of further tests on the cobalt, but expressed sadness that he may have to give up his pottery business just when he had brought it 'to great perfection'.[36] Further test results followed in March 1764, full of his usual optimism,[37, 38] and on 21st April 1764 the Society of Arts presented him with an award for having identified, at Alva, an indigenous source of cobalt.[15] The £50 prize cannot have gone far to clearing his enormous debts and soon after he moved to Bovey Tracey, near Exeter, and withdrew from the company.

Work at Alva seems not to have continued for much longer. John Williamson's progress reports cease after February 1764. The extensive work carried out in driving new levels and sinking new shafts had found only scattered traces of ore with no significant new discoveries since the cobalt vein was found in 1758. There had never been enough cobalt to justify full-scale investment in furnaces and materials necessary to manufacture zaffer or smalt on site, although some such manufacture had clearly taken place. Crisp had also attempted some manufacture, but sending ore to London to be processed into a marketable form cannot have been cheap. With no new deposits to exploit, the existing ones doubtless depleted by now, and a vital partner bankrupted, the future seemed bleak for the Cobalt Mining Company. A further blow came with the death, in 1765, of Harry Erskine.

By 1766 it seems that the Cobalt Mining Company had ceased activities on the site, for a wage receipt dated 3rd June 1766 reads:[39]

Alva 3^{d} June 1766

it is hereby Acknowledged by James Logan and John Weir that they were employd after the Mining Cobalt Company had given over working the mines at alva by John Williamson late Overseer to said Company in washing and picking the old Silver heap for Cobalt & Silver and that in the Space of fifty one days they gathered and put into seven Casks a Certain quantity of Cobalt not weighed but is Markd A: A for which trouble The Right Hon^{rble} The Lord Barjarg paid them each [illegible] one pound Eleven Shillings Stirling. and at the Same time Collected a small Box of Silver a small part of which was got at the Companys Charges but did not exeed one stone weight the rest of the Box is my Lords property and this is the truth As far as we can judge as wittness our hands day and date above writen

James Logan
John Weir

*N:B: the above Sum should be two pound Eleven shillings as appears by these Receps

Jn Williamson

Another wage receipt, dated the day before, states that the same two miners were paid £5 2s for 'one hundred and two days Work budling the old Silver heap'.[40] 'Budling' (actually spelt 'buddling') is a miners' term for washing and concentrating ore. It is surprising that the old waste was still being reworked at this late stage, even though it had been known to contain cobalt since before the Cobalt Mining Company was set up.

In March 1767 John Williamson drew up an inventory, reproduced below, of casks and boxes of cobalt remaining in the old smelt house at Alva:[34]

Note of Cobalt on hand at the Smelt house 1767

no	in Casks	St.	lb
1st wight		10	4
2d		10	12
3d		11	8
4th		11	4
5th		10	13
6th		11	-
7th		5	3
8th		4	1
9th		5	13
10th		5	-
11th		7	-
12th	This Contains Copper	6	14
13th		6	-
		105	8
	in Boxes		
No 1st		13	-
2d		9	4
3d		10	-
4th	Silver Oar	3	13
5th	part smeltd	12	9
6th		1	-
		155	2

Bearing in mind that John Williamson used a stone of 16 pounds (which becomes apparent if one adds up the sums above), this amounts to over a ton of cobalt, with lesser amounts of copper and silver. Williamson went on to explain that 'a part of the Silver in [box] number four was either Collected before the Company began the Worke or after they had stopped working at the Expense of

Lord Justice Clerke – or Lord Barjarg – out of the Clearing of the old Smeddy & Smelt house, or out of the Deads and Rubbishe of the old workes'. The first six casks were the property of the company. Williamson's comment about the remaining casks and boxes is, sadly, illegible. It is not clear if the word 'wight' in connection with the first cask is a misspelling of 'weight' or is a reference to the Isle of Wight sand brought to Alva for making zaffer (see below).

Williamson's comments again confirm that by this time the Cobalt Mining Company was inactive, a fact that dismayed one of the partners, John Stephenson of Hull, who wrote to Lord Barjarg on 5th June, 1768 saying:[41]

> It has been great concern to me to find our Cobalt Company decline to nothing. I cannot help thinking, and is what I have before observed to Mr Shirreff, that if the articles of the Copartnery had been regularly attended to there cod not have been any such loss from the Partners as appears upon the Close of the Account, & whose remissness occasiond this deficiency Seems to fall on Mr Shirreff: for on my part, whenever a Call was made, I thought it a duty in me to Send the Cashier my Share by first post, not being willing to lose my property in the lease; had a remonstrance been made to Sr Harry Erskine no doubt in his life time he wod not have Suffered So small a Sum to have remained unpaid.
>
> Mr Crisp Seems to be a misfortune, and if he has any Share of honesty in him, may in time discharge his part, but by what I find from Mr Shirreffs letters & yours he has been so craving for more Cobalt that I much doubt it, I Shall however correspond with him once more and endeavour to make something of him if I can. The Ball^ce^ to the Company is £124...6 besides £120 in the Calls made agreable to Mr Shirreffs advices but in the Acco^ts^ there are two articles of £24 & £100 – for Isle of Wight Sand, and mixed & unmixed Cobalt, that, if in our possession may in time produce the Sum of his balance abovementioned. I think if I coud have made a trip over to Alva we Shod have given the Works the utmost Trial, as every thing was ready, the Expence wod have been so trifling; I declare had there been but two partners Willing, I wod have been one of them; not giving that trial has hurt me much, as I requested Mr Shirreff to

Sir John Erskine and his wife, The Hon. Catherine, Lady Erskine.
(Sources currently unknown; photographed by Dr Ken MacKay)

Charles Erskine.
By Allan Ramsay, 1750 [detail].
(Scottish National Portrait Gallery)

James Erskine.
By Allan Ramsay, 1750 [detail].
(Scottish National Portrait Gallery)

My Dearest Life
I receiv'd Yours of the 20th & another of the 29 of Nov which were
both most acceptable, butt they had both been long by the way for
itt was the 5 of De befor I receiv'd the first, You are much mistaken
in thinking I was displeasd with you for leaveing this co... I
doe assure you I thought itt a lucky providence & tho I was in ...
for not hearing from you yett itt was easy to bear in comparison of what
terror I must have had if you had been in the danger some others of
our freinds have been in, I suppose you know all our difficultys
from better hands long Eire now & by that you may guess the
torment ... fear & terrible Horror I must be in for you & many
others if I had known your adress I had writ to you three
weeks agoe & begd of you to stay where you was till you saw
how things wold be I writ to your Brother in hops he wold learn
itt from some att Edr butt he told mee he could nott & you was
soon expected & I was so far from wishing you soon back I was
afraid to hear of your return; I pray God send a happy end
to all for I am just where I was & my hops are still very faint
that person you mention in yours not being come yett your
children are very well & all your other freinds I doe nott wish
to hear you are returnd butt when you doe pray God you
may be saffe which is the earnest wish off her who is intirly

Yours

De 10

I am better then could be expected all things considerd if you
can have any reasonable pretence to stay doe not come by any
means Mr Peck gives you his most humble service as does Aunt
B & I

[17 Jan. 1759] 19

Sir

I am sorry It has not been in my power before now to give you an account of the mineral you sent me last I was partly hindered by want of leisure & partly by an accident which befell my first trial & obliged me to begin it anew so far as I can discover it contains but a small quantity of Silver, about an ounce averdupoise in the 100 lib wt. but it is not on that account unworthy of attention It is a true Cobolt or fossil of that kind which affords the Saffer or blue for Porcelaine &c. & likewise the Smalt blue & which is sold at the Saxon mines, as dug up, at the rate of from 14 to 30 livres the 100 wt. according to its quality — yours so far as I can judge seems very good but if you find any considerable vein of it from which it may be dug up by hundreds I shall make some essays of it by making it into Smalt & Saffer upon the last of which I shall procure some trials to be made by painting some delft ware & comparing the colours with ye Saffer used here — It is observed by the miners that the silvery cobalts even of the same vein are very different in richness & where they are wrought for silver it is usual to Essay every particular heap of ore before it be sold — If therefore you meet with any peices that have a particular look you may send them to me & I shall try them — & at any rate I shall take it as a favour if you will be so good as to lay by for me any varieties in the Specimens of it that may occur because it is a fossil of the rarer kind, — The peice sent me contained a little copper but not a particle of Iron that I could observe

I am Sir yr most humble Servt

Joseph Black

Glasgow 17th Jany 1759

Above: Letter by Joseph Black reporting the first identification of cobalt at Alva, dated 17 January, 1759.
(Reproduced with the kind permission of the Trustees of the National Library of Scotland.)

Left: Letter from Lady Erskine to Sir John, dated 20 December 1715, expressing concern for his safety and suggesting it would be dangerous for him to return.
(National Archives of Scotland, GD1/44/7/13)

Alva House in the 19th century. It was considerably enlarged by the Johnstone family.
The façade is said to have been by Robert Adam (c.1790).
(Photograph by Clackmannan District Libraries)

The Silver Glen, Alva.
The mines are in the centre of the photo, but are not visible from this distance.
The hill on the left is 'The Nebit', on the right 'Wood Hill'.
The latter was planted with trees by Sir John Erskine.
(Photograph by the Author)

The main silver mine.
This is the hole from which Sir John took his fortune in silver.
(Photograph by Author)

There are many mine entrances in the Silver Glen. This one, next to a waterfall, follows a mineral vein for about 50 metres.
(Photograph by Author)

Inside a mine level in the Silver Glen.
It is barely high enough for a man to stand up in.
(Photograph by Author)

Excavation of the silver mine waste heap
by the National Museums of Scotland, in May 1994.
(Photograph by Author)

Specimen of erythrite from Alva. This is the 'peach blossom' coloured cobalt ore.
The tiny pink crystals cover an area about 5mm across. Most Alva erythrite occurs as pink stains and powdery coatings. Crystals such as these are rare.
(By kind permission of the Trustees of the National Museums of Scotland.)

Specimen of native silver from Alva.
The silver is crystallised in branching growths called dendrites. These ones are up to 30 mm long. Dilute phosphoric acid has been used to dissolve some of the enclosing matrix and expose the silver.
(Author's specimen; photograph by Author)

Porcelain plant pot, c.1770,
by William Littler of West Pans, coloured blue with cobalt from Alva; it measures 115 mm high and 146 mm wide.
(By kind permission of the
Trustees of the National Museums of Scotland.)

Communion cups of Alva silver
commissioned by James Erskine in 1767
for the local church.
(Photograph by Dr Ken MacKay)

In 1720 Sir John Erskine had some silver from his mine used to make a punch ladle
to accompany a silver punch bowl (of earlier date and not of Alva silver).
The ladle and bowl, shown here, are now the property of the Queen's Body Guard for Scotland.
(Photograph by Dr Ken MacKay)

> do it at all events. You are very kind to grant a fresh lease to any of the Partners that chuse to go on with the Adventure, I must own I am unwilling to drop the Chance after the present Efforts, I Shall be willing to join in the Same part as before, and I see by your letter Mr Crosbie & Mr Shirreff will do the Same, and I make no doubt you also will continue with us, otherwise I have no desire: Pray what is become of Mr Campble, Mr Muir & Messrs Fordyce – have they totally declined? ....

He finished by requesting 'a lump of Cobalt' for 'a Lady that is most Curious in Fossils who lives in Lancashire'. The Isle of Wight sand he mentioned was for the manufacture of zaffer and smalt and had been recommended by Nicholas Crisp, although importing it to Scotland must have been costly. It does, however, confirm that the company had been processing at least some of the ore themselves to make zaffer and/or smalt.

The proposal, mentioned in Stephenson's letter, to revive the Cobalt Mining Company seems to have been taken up, albeit half heartedly: on 11th December 1770, Alexander Shirreff wrote to Lord Barjarg bemoaning the state of the company's accounts:[42]

> I inclose the Cobalt Company Accos ballanced in my favours at the 15 June 1769 £233.12.3. There should have been a meeting, but was not, and as I despair having any, & Interest runing up, I have sent a Copy of the Accompt to each of the Partners concern'd for payment of the proportion of the Balance; now wt Interest £240.6.9. which to each Concern'd is £30. There being only Eight, as nothing will be recovored from Sr Harry Erskine or Mr Crisp.

The tone of Shirreff's letter suggests that this revival had been little more than the company's last gasp, a fact borne out by the sorry state of affairs on site as had been noted by Mr Shirreff in September 1768, just three months after John Stephenson's letter of 5th June 1768. Shirreff had observed:[43]

> When at Alva I look'd at the cobalt. There is weighed for cobalt that contains none, and is no other than stones tinged. I therefore bid John Williamson carefully goe over the whole, and pick out what was cobalt, which I believe will be so inconsiderable, as not to be worth the pains of advertising; especially as I find I can dispose of it all ....

Samples had, nevertheless, continued to be sent out to potential customers. One of these was James Watt, of steam engine fame, who at the time was involved in a pottery in Glasgow. A letter from Watt, dated 14th March 1769, described the results of his tests on the cobalt ore: '... the best gave on the Ware a very good blue but not so deep as the Dutch [German] Zaffre tho superior in Colour.'[44]

In March 1770 one John Seyfert wrote to ask if he could start work at Alva,[45] but this may have been a reference to a lease he had taken out further to the west, at 'Carnachan Burn' near Alva.[46] There is no indication of any more work in the Silver Glen at this time.

In 1773 a business partner of Alexander Shirreff went bankrupt and the same fate appears also to have befallen Mr Shirreff at about the same time,[47] but, by now, the cobalt mining was well and truly over anyway. Whatever stock of cobalt remained on site was soon disposed of. A John Bates of Lambeth requested a quantity of cobalt, 'for Tryal, about four or six hundred weight in two or three Casks' in July 1770,[35] and as late as 1775 William Littler of West Pans wrote:[48]

> I beg you will be so Kind to let me have some Cobalt I have none at all for the Kill [kiln] I am making what Ever your Lordship pleases to send for me to Edanburgh. I will Call for at your House their and hope your Lordship will Extend thy further favour wich with all others is most gratefully Acknowledged, and still hope it will be in my power some time to repay.

Whether he got the cobalt is not recorded, but the letter is significant. The Statistical Accounts record that one of the principal customers for the Alva cobalt was 'a manufacture of porcelain, that had been erected much about that time, at Prestonpans in East-Lothian'.[49] This was William Littler's West Pans works, near Musselburgh, and many pieces from there survive, decorated with 'Littler's blue' made from Alva cobalt. The tendency of the cobalt to spread during firing, noted by Nicholas Crisp, forced Littler to adopt a distinctive style in which large areas of the item were covered in blue.[50]

The Erskine association with Alva ended at about this time, when James Erskine sold Alva House and estate to the immensely wealthy John Johnstone. Even after the purchase of the Alva estate, together with estates in Selkirk and Dumfriesshire, Mr Johnstone still had a fortune of £300,000, which he had made in India.

Despite this fabulous wealth, the Johnstones showed no inclination to risk it on mining ventures and made no attempt to reopen the old workings. Instead they concentrated on improving their estate, adding large extensions to Alva House and building a stable block around 1820.

The Johnstone line ended with Miss Carolin Johnstone who died in 1920. Unfortunately she overspent so much that the sale of the house contents and the estate was not enough to clear her debts. Sadly, the huge house could not be sold and was demolished during the war.[51]

NOTES TO CHAPTER

1 Sir Henry or Harry Erskine (d. 1765), fifth baronet of Alva and Cambuskenneth, MP for Anstruther. His name was removed from the army list owing to his opposition to the Government's deployment of the Hanoverian and Hessian troops. He was later reinstated and rose to lieutenant-general. See J. Ferguson (ed.): *Letters of George Dempster to Sir Adam Ferguson, 1756-1813* (Macmillan & Co. Ltd: London, 1934).

'A truly honest man, but his views were not extensive nor his talents great.' J. Kinsley (ed.): *Alexander Carlyle: Anecdotes and Characters of the times* (Oxford University Press, 1973).

He was said to be the author of the song 'In the garb of old Gaul'. See *Scottish Notes and Queries*, IV, 3rd series (June 1926), p. 107.

2 Charles Erskine (1680-1763) merited several pages of praise by John Ramsay, who also related how Charles' eldest son, also called Charles, died when he threw himself out of a

window in London whilst delirious with a fever, 'a circumstance which was concealed from his father'. A Allardyce (ed.): *Scotland and Scotsmen in the Eighteenth Century* (Wm. Blackwood & Sons: Edinburgh and London, 1888), vol. 1, pp. 100-110.

3 NLS, E-M MSS, 5098, f. 31, 'Journal from the 28th March to the 5th April 1758'.

4 NLS, E-M MSS, 5098, f. 47, 15th January 1759.

5 Ibid., f. 52, 16th February 1759.

6 Ibid., f. 75, 18th June 1759.

7 Ibid., f. 77, 18th June 1759.

8 Ibid., f. 49, 17th January 1759.

9 Ibid., f. 28, 22nd January 1759.

10 Ibid., f. 63, 19th April 1759.

11 Ibid., f. 101, 6th December 1760.

12 Ibid., f. 122, 26th August 1761.

13 Ibid., ff. 89-90, 9th August 1760 (quoted in a letter of 10th September 1763).

14 Ibid., ff. 49-50, 30th September 1765.

15 Nicholas Crisp was a prominent figure in 18th-century porcelain manufacture and much has been written about him. For details, see J. G. V. Mallet: 'Studies in the Society's History and Archives XCVII. Nicholas Crisp, Founding Member of the Society of Arts', in the *Journal of the Royal Society of Arts*, December 1972, pp. 28-32; January 1973, pp. 92-6; and February 1973, pp. 170-4.

See also J. Turnbull: 'Scottish Cobalt and Nicholas Crisp', in *Transactions of the English Ceramic Circle* (1997), 16(2), pp. 144-51, and the references therein.

16 NLS, E-M MSS, 5135.

17 Ibid., 5098, ff. 126-7, 17th September 1761.

18 Ibid., ff. 128-9, 26th September 1761.

19 Ibid., ff. 130-1, 13th October 1761.

20 Ibid., f. 132, 1st November 1761.

21 Ibid., ff. 134-5, 5th November 1761.

22 Ibid., f. 138, 23rd November 1761.

23 Ibid., f. 178, 16th February 1762.

24 Ibid., ff. 160-1, 11th May 1762.

25 Ibid., f. 174, 28th September 1762.

26 Ibid., f. 175, 25th October 1762.

27 Ibid., f. 149, 17th February 1762.

28 Ibid., 5099, f. 1, February 1763.

29 Ibid., 5098, f. 151, 16th April 1762.

30 Ibid., 5099, ff. 7-8, 22nd March 1763.

31 Ibid., f. 9, 29 March 1763; 5099, f. 29 (undated), also mentions finding gold in assays.

32 Ibid., f. 19, 20th May 1763.

33 Ibid., f. 25, 26th or 29th September 1763.

34 Ibid., f. 60, 24th March 1767.

35 Ibid., f. 104, 5th July 1770.

36 Ibid., f. 27, 20th December 1763.

37 Ibid., f. 39, 24th March 1764.

38 Ibid., f. 41, 31st March 1764.
39 Ibid., f. 55, 3rd June 1766.
40 Ibid., f. 54, 2nd June 1766.
41 Ibid., ff. 71-3, 5th June 1768.
42 Ibid., f. 118, 11th December 1770.
43 Ibid., f. 81, 10th September 1768.
44 Ibid., f. 86, 14th March 1769.
45 Ibid., f. 94, 16th March 1770.
46 Ibid., f. 84, 1768.
47 Alexander Shirreff lived at Craigleith (now a part of Edinburgh). *In Studies in Scottish Business History*, edited by Peter L. Payne (London, 1967) p. 115, he is described thus: 'The Leith merchant Alexander Shirreff followed his father's footsteps as factor to the Earl of Hopetoun, and ultimately became a partner in a concession at Leadhills, in another in Ayrshire and in a cobalt mine at Alva: he was regarded as a leading expert in minerals, and only withdrew from these affairs when his partner at Leadhills (another Leith merchant) went bankrupt in 1773.'

The Hopetoun MSS held at Hopetoun House contain some documents connected with him. An index of these is kept at West Register House. No mention is made in the index of the mines at Alva. The silver mines they refer to are those of Hilderston near Bathgate, which were worked intermittently throughout the 17th and 18th centuries. Bundle 1795 contains papers concerning the bankruptcy of Alexander Shirreff, including an account of household furniture in Lord Hopetoun's House at Leadhills, 1770. Years are 1770-78. Bundle 1799 is a 'Scroll account settled with Mr Shirreff. 1774'.

See also Ailsa MSS, GD25, box 46, NAS, GRH, regarding his involvement in mines in Ayrshire.
48 NLS, E-M MSS. 5099, f. 147, 20th May 1775.
49 J. Duncan: 'Parish of Alva (County of Stirling)', in J. Sinclair (ed.): *The Statistical Accounts of Scotland*, IX (1796), pp. 156-160.
50 J. Turnbull: 'Scottish Cobalt and Nicholas Crisp', in *Transactions of the English Ceramic Circle*, 16 (2) (1997), pp. 144-51.

William Littler was a potter, formerly of Longton Hall, near Stoke-on-Trent, who later moved to West Pans, Musselburgh. His story is told by M. Bimson, J. Ainslie, and B. Watney in 'West Pans Story – The Scotland Manufactory', in *Transactions of the English Ceramic Circle*, VI (2), (1966), pp. 167-76, and references therein.
51 A. Swan: *Clackmannan and the Ochils. An Illustrated Architectural Guide* (Scottish Academic Press: Edinburgh, 1987), pp. 67-73.

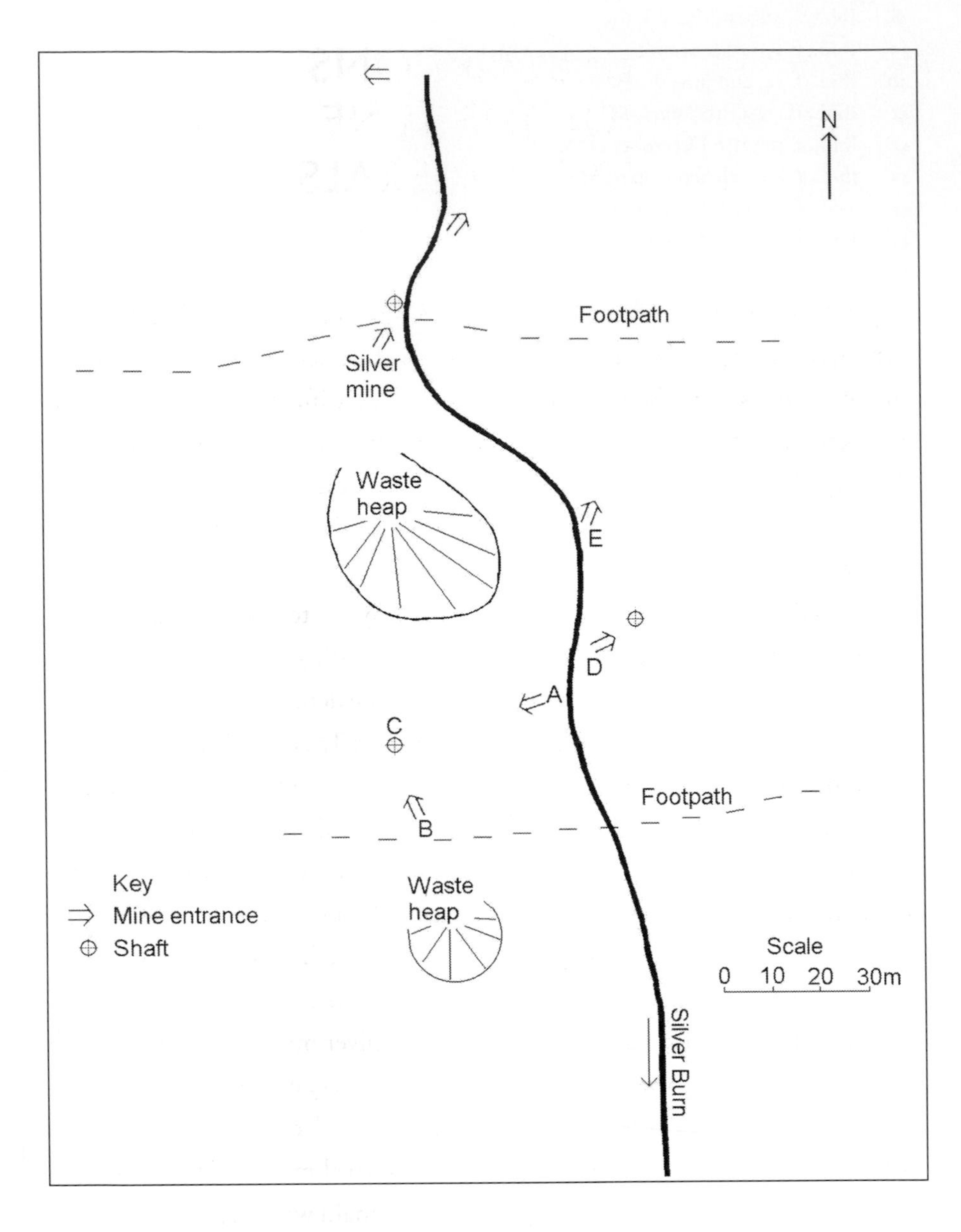

Fig. 1: Map of main workings
still visible in the Silver Glen.

CHAPTER 4

DESCRIPTIONS OF THE MINE AND MINERALS

THE story of the mines is not over. Although having been abandoned for well over two centuries, the workings have attracted considerable interest in recent years both from the historical and scientific points of view. Much new knowledge has come to light, but at the same time myths and errors have developed and become incorporated into the literature. Not least of these is the persistent, but false, notion that nearly all the silver came from the 'Silver Chamber' on the east side of the burn.

It is the aim of this chapter to dispel such myths, to describe the workings as they are at present, and to cover the recent findings of mineralogists and geologists concerning the origins and nature of the deposit.

The mines are now owned by the Woodland Trust who bought the land from the Forestry Commission in 1990. The heavily wooded hillside behind the ruined Alva House was originally planted under Sir John Erskine's ownership with both native and exotic species.[1] The stable block still stands and is privately owned. The mines are part of the Ochil Hills Woodland Park. Visitors may wander along the woodland paths at will, and one such path leads from the car park to the Silver Glen. A branch from this path takes the visitor past the cobalt mine, while the main path leads directly to the silver mine (in fact these paths are probably the old miner's tracks which owe their very existence to the mines).

The mines themselves are now mostly fenced off, but many surface remains are still to be seen in the form of shafts, level mouths, spoil heaps and trenches. The map (Fig. 1 on page 70) shows the main workings.

Sadly, the whereabouts of most of the contemporary mine plans are now unknown. The Johnstone family had a comprehensive map of Silver Glen with all the veins and workings marked.[2] Another copy was owned by Nicholas Crisp.[3] All attempts to trace these have so far drawn a blank. However, in a development that came too late for inclusion in this book, Raymond Johnstone, a descendant of the Johnstone family of Alva, generously deposited a large collection of family documents in Alloa Library. These are known to contain items relating to the mines, and it is hoped that as cataloguing of these papers proceeds interesting finds will come to light. One such, a survey of the mines made in September 1889, but with an illegible signature,[4] has been catalogued and forms the basis for Fig. 2 on page 73. Intriguingly, this concentrates on the mines on the east bank and does not even show the silver mine.

Modern surveys and excavations, together with contemporary written descriptions (which are frequently confusing, contradictory and difficult to relate to anything visible today), have enabled a partial picture of the mines to be assembled. Only in the 1980s was the true location of the main silver mine determined (by the author) and the cobalt mine tentatively identified. The locations of many of the other veins referred to in contemporary manuscripts (e.g. 'Lead vein', 'Self-open vein', 'Smiddy vein') remain unknown.

DESCRIPTION OF THE MINES

The Silver Mine

A memorandum of information provided by James Hamilton, dated 22nd August 1716, states that the silver mine was on the south side of the mountain, near the bottom, and on the west bank of a burn to the west of Sir John's house.[5] Another, undated but certainly 18th-century, document in the Erskine family papers also refers to the stream 'on ye east side' of the silver mine and states that the 'level qch was brought to serve this vein was carried alongst the South side of ye mountain westward into ye hill'. The level started by the stream or 'Rivulet'.[6] This level is most probably the level referred to in Brandshagen's plan and description of the mine.[7]

Brandshagen's plan (Fig. 3 on page 74, traced from the original for clarity) is the only contemporary plan known to have survived. The accompanying

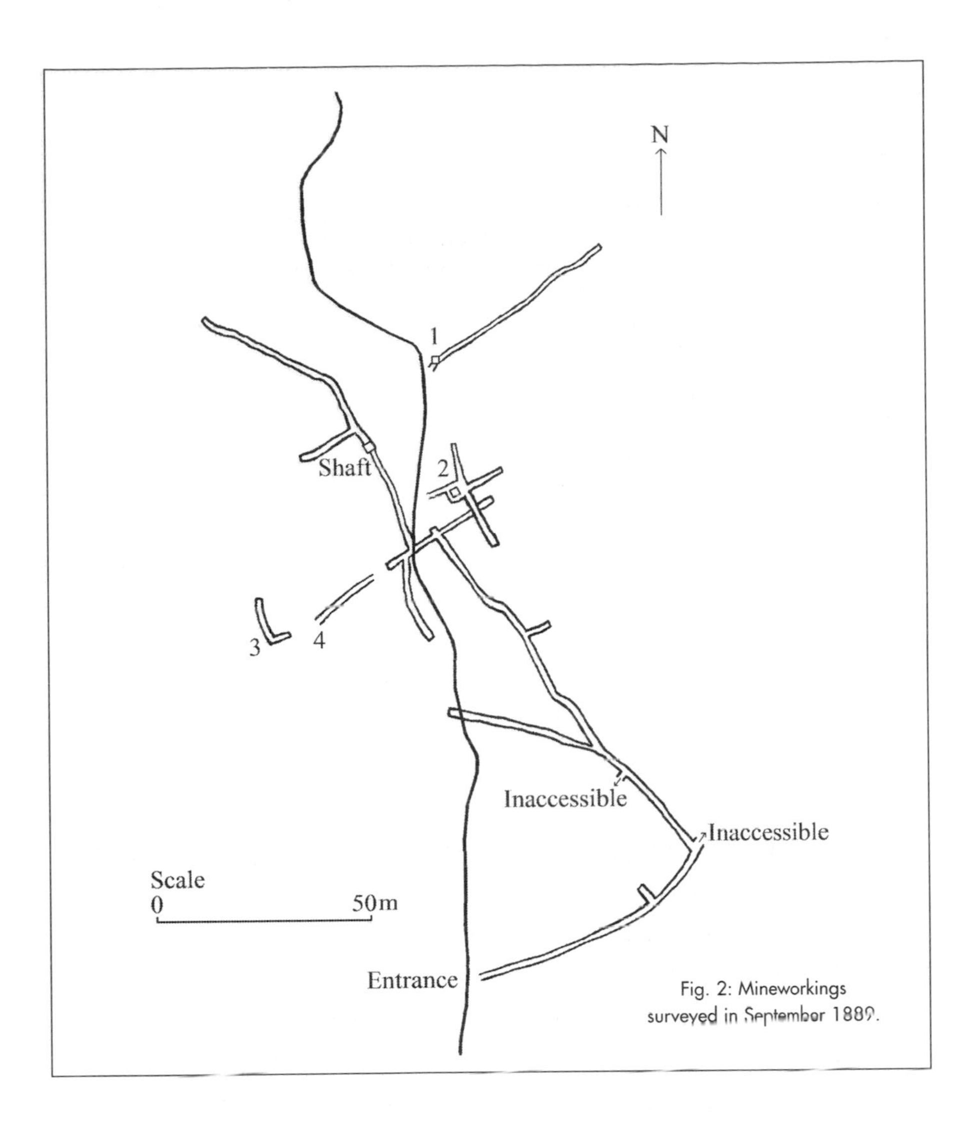

Fig. 2: Mineworkings surveyed in September 1889.

description explains that the tapering feature from X to G represents the vein, the silver-bearing part being from D to F. H is a shaft sunk on the vein and A is the main entrance level referred to above. The description reads as follows:

Description of the Mine

A is the Level to the Mine, whose Entrance is a[1]. From thence to a[2] is 22½ foot long, it is 5½ foot high and 2½ foot Wide.

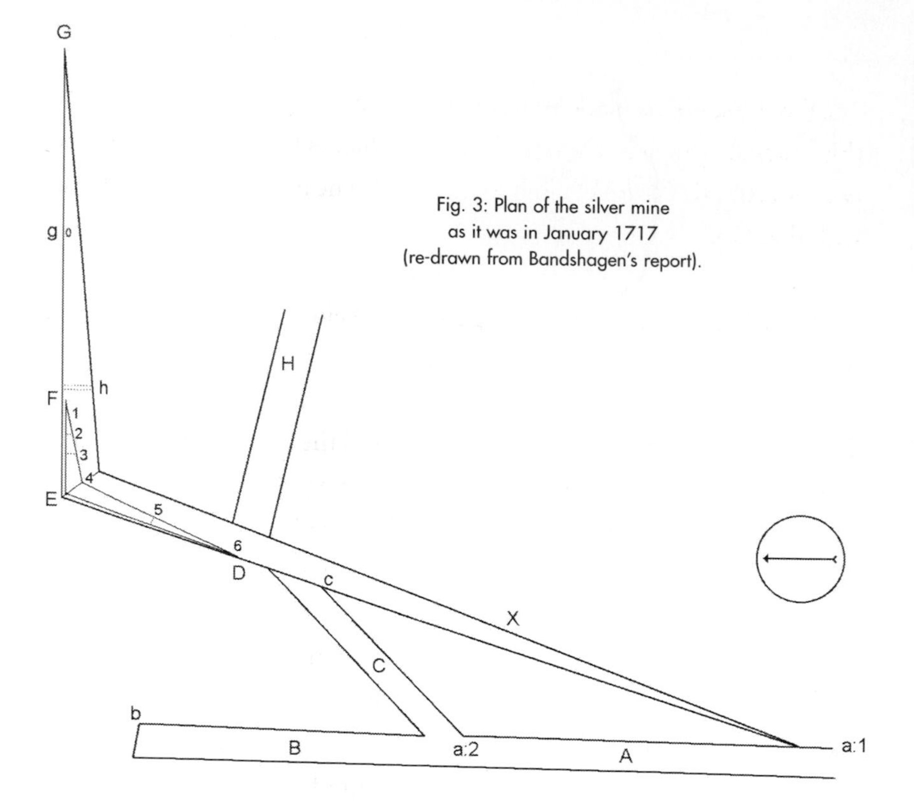

Fig. 3: Plan of the silver mine as it was in January 1717 (re-drawn from Bandshagen's report).

B is a Drift, Stretching all along equal with the Level, and is from a^2 to b 19 foot long & of the Same height and wideness with the Level. This drift is all dead Rock, and it seems they have lost the Vein in pursuing this drift, and therefore the Miniers have struck in upon the East Side at a^2 (this Section is noted in the Draught with C and runns 90 Degrees betwixt East & North) where they have found the main Vein X they have got it again at G which discovers it Self at first as a blew Clay or Marell, but very small at the Entrance of the Level A and runns in a direct Line straight along betwixt a very hard Rock 335 Degrees from West to North, this Vein continues dead w^tout any Ore till at D where the Ore appears at the bottom or Sole of the Vein 6½ foot from c: (the Vein is here wrought out from above, till it became Level to the general Level A) and as I presume to be a Drain to the Water, if they should dig deeper.

This black Vein at the beginning of it at D is but two inches broad of Ore, but the whole Vein with the Spar from Side to Side of the Vein included is one foot broad: This black Vein enlarges it Self D to E gradually, so that in this distance, which is 8 foot from D it is 15 Inches broad at E where these two parts of this black Vein join together viz[t]. The breast of the Vein F with with that Horizontal part or Sole E.D.

I come next to describe this other part of the Vein which strikes in a manner perpendicular down upon it's own part E.

This Vein appears above at the day 6 foot from the Grass, and in a loose black Earth very small above at G but encreasing gradually downwards. It is at E 2 foot 8 Inches wide from one Side of the Vein to the other (Spar and Ore included) and the Depth from G to F is 29 foot, but in this part of the Description it must be observed that from the top G to g it is all dead Spar, and at g which is 10½ foot distant from G) there appears a little String of Ore 2 Inches wide and 10 Inches downwards in length of small Value; from this little String G downwards to H is all dead Spar again, and at this H there appears almost of the same Sort of Spar but intermixed with some small Spots of Ore, and this they call here a Rider, which Rider is about one foot thick from H downwards to F and just under this Rider appears a black Vein of Ore, which Vein is like a particular Vein in the general one inclining more to the West Side of the general Vein, yea it joins almost to the Rock on the West Side, and runns from the Top to the Sole (as from F to E) 5½ foot: Above at the Top of F, the black Vein is but 3 Inches, and at No. 1. 5 Inches broad, and so the black Vein enlarges it Self gradually downwards, and at E the black vein is 15 Inches broad, as is said above, including a little Spot at the East Side, which Spot was discovered by striking off the Spar, and made the Vein somewhat broader there.

On both Sides of this black Vein, there runns all alongst downwards a dead Spar (except that in some places, it is intermixed with some small Spots of Ore) and this makes the whole Vein with Spar and Ore at E (where it appears at present broadest of all) 2 foot 8 Inches from Side to Side.

> But as has been observed this black Vein is not every where equal in Goodness and Richness of Silver; As does appear by the Assayes.
>
> H is the Shaft falling perpendicular down where the Vein D E begins viz[t] is D and this shaft is 21 foot deep.
>
> The whole Vein from the Top G downwards to E is 29 foot deep.

The plan and description came in for strong criticism by Sir John Erskine who regarded it as 'entirely nonsense' and proceeded to attack it almost line by line. In particular, Brandshagen seems to have had an unreliable compass as Sir John's remarks reveal:[8]

> ... the levell ... he makes by his compass on ye draw[ing] to runn near north it is platt west. next what he means by saying att the east side att a^1, this section is noted C and runns 90 degrees from east to north, wher they have found the main vein X, they have gott it again att G, is to me quite unknown, nor can I imagine what he designes by that part of the draught wher he would make a straight comunication twixt G & a^1 for ther never was any thing wrought ... all the rest of running straight forward in a direct line 335 degrees from west to North & c. is such an untellible nonsense, I can't make any obsevation on it other than telling it is so. But pray how will the levell serve for a drain when we dig deeper than it, as he presumes we intended.
>
> In Short if I intended to observe all the nonsense of this description, I could not better show it than by repeating his own words over again, So I wish him a good journey to London in travelling Straight along in a direct line 335 degrees from west to North.

Setting Brandshagen's dubious compass bearings aside, his description does agree well with the undated (but probably contemporaneous) one in the Erskine papers mentioned earlier. Both describe a shaft on the vein (18 feet deep according to the Erskine's document, 21 feet according to Brandshagen), both describe a lower level (A on Brandshagen's plan), and both say that a branch (C on Brandshagen's plan) led from this level to the main vein. Dimensions are also similar, the ore-bearing part of the sole (floor) of the mine (D to E on Brandshagen's plan) is given as '7 foot 2 Inches' by the Erskine's document and 8 feet by Brandshagen.

Apart from the compass bearings, Brandshagen's report is probably quite accurate. However, the same cannot be said of James Hamilton's descriptions. In his affidavit to the Lord Mayor of London,[9] he alluded to two veins of ore, with more details being included in his memorandum of 22nd August 1716:[5]

> The Mine is just opened, within about two fathoms or 2½ from the grass wch grows on ye surface of ye earth. In it are two veins of ore running horizontally the one almost three foot above the other, the upper vein about 22 inches broad from top to bottom & about 18 inches wide, the other about 14 inches broad or deep & about the same wideness with the former.

This is flatly contradicted by the aforementioned document from the Erskine papers which states: 'There is but one vein going perpendicular, a little heading to ye east. No horizontal veins in any part of ye work.' It is clear also from Brandshagen's plan that there was just one vertical vein in the silver mine, and his report contains the line: 'That as to what Beds of Silver or Copper Ore appear in the Mine, neither the Hamiltons nor I could observe nor find any, nor no other Veins more than what is already Represented.' Regarding this remark, Sir John commented that it was 'entirely nonsense and not near so explicitt as to the contradicting the two horizontal veins sworn to by J. Hamilton as it ought to be'.[10]

A document dated 1753, entitled 'Memorandum of Facts concerning the Mines of Alva wrought by S. Jo. E. taken from the mouth of one of the miners many years after', gives the dimensions of the silver orebody as four fathoms vertically and eleven feet horizontally in a vein up to three feet thick.[11]

There is only one working visible today, on the west bank of the burn, that could fit these descriptions. It is located directly under the point where the main footpath crosses the stream (Fig. 1 on page 70). The entrance to the level was probably at the bottom of the waterfall but is now obliterated. However, when the stream is low a trickle of water can be seen emanating from this point.

The accessible part of the mine consists of a section stoped to the surface (a stope is a section of vein that has been removed) on a vein trending roughly north-east/south-west. Two levels have been driven from the northern part of the stope, the upper is blocked at the bottom of an infilled shaft (the top of which can be seen by the footpath). The lower one follows the vein for about 18 metres before terminating. There appears also to be a blocked or collapsed level driven southward from this stope.

Immediately above the mine, and on the east side of the stream, some open-cast trenching marks the outcrop of the vein, and a short distance to the north-east is a level of about nine metres length on the same vein.

A path leads down from the silver mine directly to its spoil heap – the largest in the glen. It is this heap that the miners reworked in the 1760s for cobalt, but they appear not to have worked it all. Whilst the bulk of it is quite barren of ore minerals, a small area of a few tens of square metres at the top and around the east side was found by mineral collectors in the 1980s to contain approximately a half to one ounce of silver in every ton. Most of this was as tiny fragments and individual crystals rarely more than two or three millimetres across, but a few exceptional pieces of ore up to several centimetres across turned up, together with much erythrite and other rare minerals. An Irish halfpenny, dated 1686, found at a depth of about half a metre below the summit of the dump (and now in Alloa Museum), confirms the antiquity of this mine waste. Further archaeological evidence comes from a clay-pipe bowl found on the east flank of the tip in 1996. This was identified from its maker's mark as being the work of either John or James Paterson of Stirling, both of whom were known to have been active in 1685.[12]

The Cobalt Mine

While several detailed descriptions of the silver mine have survived, there is no comparable description of the cobalt mine still in existence (to the best of the author's knowledge). As mentioned in the previous chapter, there are tantalising references to it in John Williamson's reports to his bosses Alexander Shirreff and James Erskine, and in other documents. However, about the dimensions of the orebody, its exact location and the tonnage produced, we know virtually nothing. We know that the cobalt vein was intersected by 'Spar vein' (the intersection being found on driving westwards along the latter) and at the intersection the cobalt vein 'points near 12 degrees west of north',[13] which is different to the roughly north-east/south-west trend of the silver vein. We also know that Spar vein points due west and that a cross-cut driven from it led to another vein and then also to Cobalt vein.[14] This would suggest that the cobalt vein lay on the west side of the stream. It is also reported by the Statistical Accounts (see note 2 at the end of this chapter) that the cobalt was found during the digging of a drainage level for the silver mine which would place it below the silver mine.

There is only one group of workings visible today on the west side of the stream, below the silver mine. These are the workings marked 'A', 'B' and 'C' on Fig. 1 on page 70. Of these, only 'B' is shown on the plan of 1889 (mine working '3' on Fig. 2 on page 73). Level A points westward along a vein and could conceivably be the level along Spar vein. Level B points towards the Silver vein and could well have been an attempt to drain it. If so it was never completed, as it stops far short of the silver mine workings. Immediately to the east of this level is an open-cast working or trench, now filled with rubble. The eastern end of this may be a blocked mine entrance. If this represents a vein, then it would have been discovered just as work began on level B. This would be consistent with the story in the Statistical Accounts of the cobalt vein being found just as the men began work on the drainage level for the silver mine. This feature is shown on the plan of 1889 as a short length of tunnel (feature '4' on Fig. 2) and is clearly aligned with deeper workings, presumably along a vein at that point.

Of course all this is rather speculative and only a proper excavation centred on workings A, B and C will resolve it. Perhaps significantly, the earliest 1:2500 Ordnance Survey map (dated 1862) has the area around B marked as 'Cobalt mine'. For the silver mine, and also mine D, the Ordnance Survey map says 'Silver and Cobalt Mine'.

Mine A can be followed for about 12 metres before it is blocked by rubble. Surprisingly, it does not appear to connect with shaft C, which is sunk on the same vein and is directly in line with level A. The bottom of this shaft may be accessed by level B, but is partly blocked with debris fallen in from above consisting of soil, rubble, sheep bones, bottles and other rubbish. Some digging in this muck encountered solid rock and it did not appear to continue down to the level of A. The fact that it floods in wet weather rather than draining away also points towards the absence of any connection with A, but only completely clearing it, or clearing level A, will resolve this.

Where the waste went from the cobalt mine is also a mystery. The small waste heap from level B is only barren rock. It may have been back-filled underground, or used to fill up dangerous workings at a later date. Whatever the case, its absence is bad news for any mineralogists seeking specimens on surface.

The Legend of the Silver Chamber

For a long time it was popularly believed that the mine marked 'D' on Fig. 1 ('2' on Fig. 2) was the main silver mine. The level opens into a 6 x 4 metre chamber with a shaft to the surface. The chamber is at the intersection of two veins, one roughly east-west, the other nearly north-south, and three levels lead off from the chamber for up to twelve metres along these two veins. In addition a shaft has been sunk at the south-west side of the chamber for at least seven metres (44 feet according to the plan of 1889).[4]

The tradition that this was the silver mine seems to have originated with G. V. Wilson's report, 'The lead, zinc, copper and nickel ores of Scotland', published in 1921.[15] This gives an incomplete description of the workings in just one paragraph as follows:

> Three levels have been driven along the veins, and two shafts have been sunk. The one on the east side of the burn is connected with the levels, but the one on the west is only connected with shallow workings. A winze has also been sunk from the middle adit for a depth of 7fms, to another level from which another winze 11fms. deep has been sunk to the low adit. The native silver is said to have been obtained near the position of the top winze.

This was interpreted in the *Memoirs of the Geological Survey* (1970)[16] as referring to the chamber on the east bank, it being assumed that the shaft (or winze) in the chamber connected with lower levels (which it probably does). Once this was in print, other publications repeated the error. The Clackmannanshire Field Studies Society's booklet 'Mines and Minerals of the Ochils' (1974)[17] elaborated the myth further and a team of volunteers tried to clear the debris from the shaft in the hope of reaching lower levels, including the long-lost cobalt mine, but were unable to complete the task.

In 1978 the British Geological Survey drilled a borehole to intersect the veins in this chamber at depth. Their report is on sale at their Edinburgh headquarters (Murchison House, West Mains Road) and includes a map with 'Silver Chamber' marked, and diagrams showing how they thought the ore may have been deposited in a swelling at the intersection of the two veins in the chamber.[18] This is not an unreasonable proposition as such intersections sometimes do contain rich deposits, but in this case it was based on the false premise that the chamber was the silver mine.

Even in scientific papers as late as 1988 the error was still being repeated.[19] The most recent (1991) manifestation of the silver chamber myth was a display in a nearby visitor centre (then located in the Alva House stables) which simply repeated the whole tale.

This story shows how damaging a single error can be once it becomes entrenched in the literature, and how it grows with time. It is surprising this one lasted so long. Even to a casual observer it is obvious the spoil heap could not have come from the chamber. The chamber is too small to have yielded the wealth Sir John obtained and the large working on the west bank must have once contained a substantial mass of ore. One wonders what Sir John Erskine would have said.

Other Workings

There are several other workings in the glen, mostly just short trial levels or shallow trenches. These are described in references 16 and 17. Mine E on Fig. 1 ('1' on Fig. 2) is the longest level still open in the glen. It follows a calcite vein for about 50 metres. About seven metres from the entrance is a flooded shaft. This is not the only hazard however. The veins in the Silver Glen carry traces of uranium. Decay of this generates the radioactive gas radon – thought to be the second most important cause of lung cancer in Britain. This invisible, odourless gas accumulates in any poorly ventilated space – the basements of houses for example. Mine workings are particularly good at trapping the gas as it seeps from the rocks, and traces have been detected in all the workings on Fig. 1. Mine E gave the highest reading of them all.[20]

Workings not previously described in the glen are a 12-metre adit pointing west, together with some trenches on the west bank at a confluence further up the glen. A very substantial trench south-west of this, parallel to and just west of a fence, goes straight down the hill, almost to the silver mine.

More significantly, there is a collapsed level, much lower down the glen, driven eastwards along the Ochil fault. Its presence is betrayed by a gully with water flowing out, the remains of a bridge and a flat area of spoil on the opposite bank. Eastwards of this, and in line with it, is a collapsed or filled up ventilation shaft, now just a circular depression in the woods. Roughly along a line between this and mine D are the remains of another old shaft. Doubtless this is the extensive working surveyed in 1889 and shown in Fig. 2. The two shafts correspond with the points marked as 'inaccessible' on the plan and seem to be slightly offset from the main level. This working represents an attempt by the miners to intersect the veins at a much greater depth. Indeed it may well connect with lower levels of mine D in Fig. 1 ('2' on Fig. 2) as the shaft in D never floods even though it is below the level of the stream and often has a trickle of water down its side. Such a connection via internal shafts (winzes) and an intermediate level is described in the extract from Wilson's memoir above.[15]

In Fig. 2 this level reaches a 'T-junction' below mine '2'. Most likely this represents the intersection with a vein, perhaps the cobalt vein, which has then been followed in both directions. A branch continues towards the silver vein, and probably reaches it, although the absence of any working along the silver vein would suggest that the vein was barren at this depth. Were the entrance to this level to be dug out it would render accessible hundreds of metres of passages. The considerable extent of this lost mine working demonstrates how determined the Cobalt Mining Company was to find further reserves of ore.

There are many veins and workings described in various contemporary reports, but without contemporary maps or plans it is difficult or impossible to relate them to anything visible today. Some seem to have become totally obliterated.

Tonnage

It is difficult to estimate how much silver was produced. Some 40 tons of ore were dug out during the 1715 rebellion and buried again in casks. If this was as rich as the assays suggest then this may represent as much as two or three tons of silver. Lady Erskine's words, 'all of a sudden there was no more valuable thing to be got in that place … & so [the mine] was shut up' on June 11th, 1716,[21] seem surprising in view of the fact that there was still a considerable quantity of extremely rich ore in place when Brandshagen inspected the mine. Indeed, judging by the dimensions of the ore body given by one of the miners – four fathoms vertically and eleven feet horizontally – it seems likely that the bulk of the deposit was still there. The size of the spoil heap would also suggest that more than just 40 tons of ore were mined.

Apart from mention of 270 stone and, a little later, 180 stone weight of ore being raised around 1719, records of the amount of ore produced after the rebellion seem not to have survived. The only guide to the total output comes from the Statistical Accounts[2] which suggest that Sir John's profit was £40,000 to

£50,000. As the price of silver was fixed at 5s. 2d. per troy ounce, this would represent 155,000 to 194,000 Toz or approximately five to six tons of silver. This estimate was, however, conjecture. Nevertheless the Accounts do state that 'it has been credibly affirmed that ore was produced to about the value of L. 4000 *per* week' over a 13 or 14-week period. This alone would amount to between six and seven tons of silver and does not include that mined before or after this period. Whether this glorious 13 or 14-week bonanza was during the rebellion, or after (perhaps in the 1720s), is something we may never know. The amount stolen by the miners was also said to have been considerable.

The output of cobalt is even harder to estimate. As described in the previous chapter, probably about a ton was sent to Mr Crisp in just one shipment in 1763. How many other shipments were made is not known. Another ton was still in storage on site in 1767.

The Minerals

No account of the silver mine would be complete without mention of the extraordinary diversity of common and rare minerals found there. To date (2005) some 35 mineral species have been positively identified from Alva. These can be categorised as follows:

native elements: silver
arsenides: clinosafflorite, maucherite, niccolite, rammelsbergite
selenides: tiemannite, unnamed silver/bismuth selenide
sulphides: acanthite, galena, bornite, chalcocite, chalcopyrite, covellite, digenite, djurleite, marcasite, pyrrhotite, pyrite, sphalerite
oxides: uraninite, hematite, goethite, quartz
arsenates: annabergite, conichalcite, erythrite, picropharmacolite, tyrolite
carbonates: malachite, azurite, dolomite, calcite
miscellaneous: barite, pyrobitumen, kaolinite

It should be noted that many of these occur only in trace amounts and require specialist equipment, such as electron microprobes, to detect them. In addition to these, there are several minerals said to have been found at Alva but which have never been properly substantiated. These are native bismuth, argentite and smaltite recorded by Heddle,[22] and argentite (again), and arsenopyrite recorded by Wilson[15] and copied by Francis, *et al.*[16] It is possible the smaltite is a mistaken identification of clinosafflorite, which it closely resembles. Similarly, argentite is difficult to distinguish from acanthite. The bismuth is possible, but must be very rare as none has been found in modern times. There is also mention in the Erskine papers of an ore suspected of containing it.[23] Arsenopyrite remains completely unconfirmed with no known specimens anywhere.

The silver occurred almost entirely in the metallic state – i.e. as native silver. It was usually very well crystallised as dendrites, that is in branching forms reminiscent of ferns or Christmas trees. These dendrites attained lengths of several centimetres and were found embedded in a matrix of dolomite and/or clinosafflorite or, less often, barite or calcite (the 'spar' of the miners). Less commonly the silver formed irregularly branching masses or small cubic crystals up to one millimetre often in groups and strings, again embedded in dolomite. Two 19th-century descriptions of the silver refer to strings of octahedra forming the axes of a larger octahedron.[22,24] Many dendrites on modern specimens are terminated by half-octahadra and, when not fully exposed, may resemble strings of octahedra. Some beautiful specimens of the silver were pictured in a recent article in the *Mineralogical Record.*[25]

The ore was extremely rich. Brandshagen's report gives the results of seven assays of ore taken from the points indicated on his plan. Converted to percentage silver in the crude ore, the results were 2.4 and 3.5 % for two samples from point 1, and 3.4, 1.1, 5.8, 0.1 and 0.5 % silver for samples from points 2 to 6 respectively on Fig. 3. An assay result in James Hamilton's affidavit gives 15½ pennyweight silver from 16 oz of untreated ore (4.8 % silver). Other assays by Sir Isaac Newton gave similar results (up to 22 dwt silver per pound of ore, or 7.5 % silver).[26] His samples contained little or no copper and no gold. The assay given in the Statistical Accounts (12 oz of silver in 14 oz of ore, or 86 % silver) is so much richer than the other assays that it may have been carried out on processed material, washed free of adhering minerals, and is not representative of the crude ore.

Modern analyses of the silver show it to contain 12-16 % mercury, together with approximately 0.7 % bismuth and 0.3 % antimony, but again no copper or gold.[27] Thus the native silver is about 83-87 % pure, which is strikingly similar to the Statistical Accounts' figure.

The main cobalt ore is the cobalt arsenide mineral clinosafflorite. This occurs as shiny grey to black fragments and grains (the 'metallick particles' referred to by Crisp) and is normally coated with pink erythrite (the ore with the colour of 'peach blossom'). Clinosafflorite is easily mistaken for the closely related mineral safflorite, and initially this is what happened when the first modern specimen was found in 1980. Subsequent work, however, has shown that all the primary cobalt arsenide ore from the silver mine (including the first piece thought to be safflorite) so far examined is, in fact, clinosafflorite. This caused a little excitement amongst Scottish mineralogists as clinosafflorite is an exceedingly rare mineral known from only a handful of locations around the world (silver and cobalt mines in Ontario, Morocco and Sweden). This and the other minerals are described more fully in recent papers.[25,27]

Anyone hoping to find silver in the glen nowadays is likely to be disappointed. It was never common, and after years of thorough searching by experienced collectors and mineralogists it is scarcer than ever. The small part of the mine dump left untouched by the miners has been completely dug and sifted through – right down to the original soil level (generally less than a metre deep). In May 1994 the National Museums Scotland used a mechanical digger to excavate the site.[28] The richest portions of the waste were washed with water pumped from the stream. A week of this yielded just 20 to 30 fragments of silver ore. A metal detector search found a shilling dated 1839, but no native silver.

In the process of these excavations some clues came to light regarding the working of the mine. The presence of shot-holes drilled by hand showed the mine was worked with gunpowder (blasting was mentioned in James Hamilton's affidavit), and from the relatively undamaged state of individual, detached dendrites and crystals of silver it is clear that the ore was crushed rather than ground. (Experiments by the author have found that grinding causes silver dendrites to roll up into little cylinders, crushing only flattens them slightly). An undated, but probably 18th century, note in the Paul MSS describes the working of native silver mines in Norway.[29] There the ore was stamped to a powder and the lightweight waste washed away with water leaving the denser silver behind. The same appears to have been carried out at Alva.

Although the chances of finding silver at Alva are now remote, traces of cobalt in the form of thin coatings and stains of pink erythrite ('stones tinged' as Alexander Shirreff put it) can sometimes be found with diligent search. These do occasionally contain silver and looking for them is probably the best strategy for silver seekers. They are often accompanied by light blue stains of the copper calcium arsenate mineral tyrolite. Barite remains abundant and large lumps can easily be found.

Much of the material collected from Alva in the last few years now resides in universities and museums. National Museums Scotland, Manchester Museum and the University of Edinburgh's geology department hold the most, including many specimens of the cobalt and nickel arsenides and native silver. Smaller collections including silver are held by Alloa Museum, the British Geological Survey and the Scottish Mineral and Lapidary Club. Dozens of small specimens are widely dispersed in private collections.

Sadly, very few specimens from the 18th century are known to have survived to the present day. A search around most of the main British museums has revealed only a handful of specimens. The best of these is a magnificent specimen of native silver (specimen M953) in the Hunterian Museum in Glasgow. Details of other historic specimens are given in the notes and references.[30]

Other relics

In 1767 Lord Barjarg, ignorant of their value to the science of mineralogy, had some samples of silver ore that had been preserved from Sir John's days melted down to make a pair of communion cups for the local church. On these is engraved in Latin: 'Sacris in Ecclesia S. Servani apud Alveth, A.D. 1767, ex argento indigena. D.D.C. Jacobus Erskine.' Roughly translated, this means: 'Dedicated to the Church of St Serf's in Alveth [the old name for Alva], A.D. 1767, made from indigenous silver. Consecrated to the Divine God. James Erskine.'

There is also a punch ladle of Alva silver bearing the date 'Decem. 20, 1720' made to accompany a punch bowl of earlier date (and not of Alva silver). It cost 8s. 2d. to make. Bowl and ladle are now the property of the Queen's Body Guard for Scotland. Although the Erskine family are said to have had a number of items made of their own silver for their personal use, the cups and ladle are the only ones with adequate documentation to prove their origin.

Origin of the deposit

The geological history of the Ochils has been long and complex and it is not possible to give more than the briefest summary here. More details can be found in note 16 to this chapter. The Ochils are mostly volcanic in origin, being composed of lavas, agglomerates and tuffs erupted some 400 million years ago in the Devonian period. Tectonic movements of the earth's crust caused the central lowlands of Scotland to begin to subside later in the Devonian and throughout the Carboniferous period. Because subsidence was much greater to the south than to the north, a large crack, or fault, developed with the rocks to the south subsiding rapidly. This fault is what we now call the Ochil fault; it runs along the foot of the southern flank of the Ochils and is responsible for the dramatic change in topography.

Thick accumulations of sediments built up on the subsiding land south of the fault and great forests grew and died, their buried remains forming rich coal seams. Molten rock under great pressure was forced up the line of weakness formed by the fault and formed a line of quartz-dolerite dykes. Using radio-isotope methods these have been dated to about 295 million years in age, near the end of the Carboniferous period and about the same age as the mineralisation. Similar dykes of the same age are associated with barite deposits at Muirshiels in Renfrewshire and, more relevantly, with a small native silver deposit with nickel, cobalt, arsenic and lead, similar in many ways to that at Alva, which was discovered in 1606 near Bathgate. (The story of this mine – the Hilderston silver mine – is almost as remarkable a tale as that of Alva and is well worth reading).[31]

Researchers at the University of Strathclyde, including an Iraqi student Rafaa Zair Jassim who completed a MSc thesis on the deposit in 1979,[32] suggested that the heat from the quartz-dolerite intrusions caused water in the cracks and pores of the surrounding rocks to circulate in a convective system. This water would be extremely hot, well over 100°C, but unable to boil because it was in a confined space – a kind of pressure cooker effect. Under these conditions water will tend to leach out from the surrounding rocks elements normally present only in minute trace amounts – copper, lead, barium, silver, etc. – and concentrate them in solution.

As these hydrothermal solutions, as they are called, rose up the faults and fissures in the rock towards the surface, the temperature and pressure fell. As things are generally less soluble in cold water than in hot, so the dissolved substances began to crystallise on the sides of the fissures, gradually filling them up with minerals, thus forming mineral veins. In the case of the mineral barite, this was chemically precipitated when the solutions carrying barium chloride mingled and reacted with sulphate in groundwater seeping down from above.[33]

Differences in temperature, pressure, nature of the rocks, pH and composition of the fluids, etc., all affect the metals and minerals deposited. Copper, barium and lead are easily concentrated under a wide range of conditions and so are widespread, but the conditions required for silver are much more unusual and demanding and, consequently, less well understood. Accordingly, in only a few localised areas of the circulating system of solutions were the conditions just right for silver to be deposited, and it was one such spot that Sir John Erskine's miners discovered. The conditions that concentrate and deposit native silver also happen to be the same for cobalt, nickel and arsenic, and where native silver occurs these other metals frequently accompany it.

More recent research, and comparison with other deposits (particularly those at Cobalt, Ontario, which are very similar to Alva), has tended to shift the emphasis away from the dolerite dykes and more towards the rifting and subsidence that formed the central lowlands of Scotland. In this scenario the dolerite dykes may have helped form some of the cracks and fissures for mineralising fluids to pass through, but are otherwise quite incidental, the hot solutions being mobilised by the deep faulting and the subsidence to the south.[27]

Future prospects

Mineralisation is widespread in the Ochils, and ores of copper, iron and lead are present in addition to those of cobalt and silver. These have been mined at many locations, mostly on a very small scale. The various workings are described by Dickie and Forster.[17] The most detailed account of their history is that by Harrison.[34]

At present the main excitement in the Ochils concerns the discovery of gold in streams, particularly around Glendevon to the east.[35] The sources have yet to be located, but it is thought that this gold mineralisation, associated with the mercury ore cinnabar, is not related to the silver, barite and base-metal veins along the southern flank of the western end of the Ochils. The barite veins at Blairlogie have also received some attention, but their high content of impurities makes them unattractive.

Most of the veins in the Silver Glen were found to contain small amounts of 'silver ore'. Daiglen Burn mine (grid ref. NS 91069834), above Tillicoultry, was worked for copper, but some native silver and cobalt ore were also found. There was also an 'Airthrey Silver Mine', but its location is unknown. The site tentatively identified as being this mine[16,17] is a small trial on an almost barren vein which could not have yielded the 12½ tons of ore, valued at £60 per ton, said by the Statistical Accounts to have been raised. It may actually have been the Airthrey copper mine.[34]

Carnaughton Glen mine (NS 87819754) was said to have been worked for silver, but is no more than a trial. It probably represents John Seyfert's activities mentioned in the previous chapter. Traces of silver are said, by the Statistical Accounts, to occur in a vein in the Burn of Care, near Castle Campbell, above Dollar. A piece of native silver 'the largeness of a bean, malleable' was found in a lump of 'spar' in 'the Westward Burn' (probably Alva Glen) in the 18th century.[11] Recently, amateur gold-panners prospecting in streams east of Glendevon have found fragments of silver very similar in appearance and composition to that from Alva.

It is clear from these scattered occurrences that mineralising fluids carrying silver were circulating over a large area of the Ochils from Airthrey in the west to beyond Glendevon in the east. Furthermore, deposits of this type are not normally one-offs. Where there is one native silver vein there are often dozens of others. The discovery of the cobalt and silver veins at Alva was largely due to luck, and the fact that they were close to the surface. It is almost certain that many others are hidden at depth, or buried beneath the thick blanket of glacial debris that covers so much of the hills. One is reminded of Charles Erskine's remark, quoted earlier, that, 'If one could turn over the Ochils like a beehive, something might be got worth while'.

NOTES TO CHAPTER

1 R. Paul: 'Alva House Two Hundred Years Ago: part II', in *The Hillfoots Record*, part I (27th March 1901), p. 3; and part II (10th April 1901), p. 3.

2 J. Duncan: 'Parish of Alva (County of Stirling)', in J. Sinclair (ed.): *The Statistical Accounts of Scotland IX* (1796), pp. 156-60.

3 Mentioned in NLS, E-M MSS, 5098 f. 157-8 (4th May 1762), and f. 180 (undated).

4 Unknown (signature illegible): *Plan of workings in Silver Glen, Alva*, Johnstone MSS, Alloa Library (1889), PD239/204/3.

5 NA, MINT 19/3/233, 22nd August 1716.

6 NLS, E-M MSS, 5099 f. 170-1 (undated).

7 NA, T64/235 (original copy, 1717). Another, contemporary, copy is in the Lauderdale papers, NAS, WRH, RH4/69/26/6.

8 NLS, Paul MSS, 5160 f. 10-11 (undated).

9 Ibid., f. 5, 3rd July 1716.

10 Ibid., f. 12-13, undated.

11 NLS, E-M MSS, 5098 f. 26-7, 1753.

12 Dr David Caldwell, National Museums of Scotland, in personal communication. How long the Patersons remained in business is not known, but the presence of one of their pipes in a waste heap dated from historical sources to no earlier than 1715 suggests that one or both were still active at this time. Unlike coins, which could circulate for decades, clay-pipes tended to be used for a short time, then discarded. For more information, see P. Davey (ed.):

'The Archaeology of the Clay Tobacco Pipe X Scotland', in *British Archaeological Reports*, British Series 178, (1987), particularly the section on 'Early Pipemaking in Stirling' by D. B. Gallagher, pp. 165-6.

13 NLS, E-M MSS, 5099, f. 30, 16th January 1764.

14 Ibid., f. 33, 10th February 1764.

15 G. V. Wilson: 'The Lead, Zinc, Copper and Nickel Ores of Scotland', *Memoirs of the Geological Survey* (HMSO: Edinburgh 1921), pp. 143-4.

16 E. H. Francis, I. H. Forsyth, W. A. Read and M. Armstrong: 'The geology of the Stirling district', *Memoirs of the Geological Survey, G.B. – Scotland* (HMSO, Edinburgh 1970).

17 D. M. Dickie and C. W. Forster: 'Mines and Minerals of the Ochils', published by the Clackmannanshire Field Studies Society (1974).

18 I. H. S. Hall, M. J. Gallagher, B. R. H. Skilton and C. E. Johnson: 'Investigation of polymetallic mineralisation in Lower Devonian volcanics near Alva, central Scotland', *British Geological Survey, Mineral Reconnaissance Programme Report*, no. 53 (1982).

19 J. Parnell: 'Mercury and silver-bismuth selenides at Alva, Scotland', *Mineralogical Magazine* (1988), 52, pp. 719-20.

20 Detected by R. J. Cole, Inspector of Mines in April 1990, and reported to the Woodland Trust. His letter said that the mines are unsuitable for tourism purposes and recommended fencing off the mine entrances. If this was to be done in a way that permanently sealed the mines it would be a tragedy, as underground excavation, by properly equipped and experienced persons, particularly in the cobalt mine area, would be immensely interesting both historically and mineralogically.

21 NAS, GD1/44/7, 11th June 1716.

22 M. F. Heddle: *Mineralogy of Scotland*, vol. 1 (Edinburgh 1901).

23 NLS, E-M MSS, 5098 f. 130-31, 13th October 1761.

24 J. Sowerby: *British Mineralogy*, vol. 3, pl. 327, pp. 45-6 (1809).

25 S. Moreton: 'The Alva Silver Mine', *Mineralogical Record* 27 (November to December 1996), pp. 405-14.

26 NA, MINT 19/3/268 (undated).

27 S. Moreton, P. Aspen, D. I. Green and S. Ingram: 'The silver and cobalt mineralisation near Alva, Central Region, Scotland', *Journal of the Russell Society* 7(1) (1998), pp. 23-30.

28 B. Jackson: 'Mineral rescue collecting at the Alva Silver Mines', *Forth Naturalist and Historian* 17 (1994), pp. 3-4.

29 NLS, Paul MSS, 5160 f. 20-1 (undated, anon).

30 18th-century museum specimens of Alva ores. At the time of writing (2005) only those specimens in Glasgow, Hunterian and Liverpool museums have actually been seen by the author.

Smith Museum, Stirling

A specimen of silver ore was once owned by the museum but it disappeared sometime in the early 1970s.

Glasgow Museums and Art Galleries

Specimen G1977-115-4 (from the collection of Sir John St Aubyn, 1758-1839) is a tennis-ball sized lump of black/dark grey cobalt arsenide (supposedly smaltite but this needs to be confirmed) with much erythrite and limonitic material. It is very similar to specimens found in the mine dump at Alva.

Specimen 96-65 AJV (smaltite, erythrite and sphalerite from the collection of D. C. Glen) looks very different to anything found in the mine dump. In addition, the large amount of sphalerite it contains is unusual for Alva where sphalerite is extremely rare.

Specimens G1977-115-5 (cobaltite, smaltite and skutterudite) and 02-161DB (stephanite) look totally unlike Alva material and may not be authentic.

Hunterian Museum Glasgow

At least three native silver specimens and one grey cobalt ore specimen from Alva are listed in the old catalogues of Dr Hunter's collection. He bought a specimen of 'Native Silver with Kobalt from Alva' for a guinea on 20 April 1778 from a Peter Woulfe. There are descriptions of 'Silver Crystallized in large irregular crystals, in the interstices of a load of Cobalt – From Alva Scotland', 'Silver in a dendritical form, running through the Cobalt Ore – From the same load' and 'Netted Silver Ore, the interstices of which are filled with calcareous Spar from Alva'.

Relating these descriptions to actual specimens in the collection is not easy, but the last one above is specimen no. M953, a stunning mass of absolutely typical dendritic Alva silver about the size of a hen's egg. Most of the matrix (calcite) has been dissolved away with acid to reveal a mass of coarse dendrites up to three centimetres long.

Cliffe Castle, Keighley, West Yorkshire

This museum houses the collection of Joseph Dawson (1740-1813). His catalogue lists two specimens (nos. 2161 and 2162) of 'Grey Cobalt Ore' and one (no. 2163) of 'Black Cobalt Ore' from Scotland.

National Museum Liverpool

There is a specimen in Liverpool museum purporting to be silver from Alva. It consists of a few silvery specks in quartzose material which looks unlike anything from Alva (where quartz is very rare). An old record says the specimen contained barite (which would make sense for Alva). It seems likely that specimens have become mixed up at some time in the past and the one currently labeled as being from Alva is not. The fact that most of the museum's collections were destroyed by bombing in the Second World War may not have helped.

British Museum (Natural History), London

For the following descriptions I am indebted to Nick Carruth of Cornwall:

Chalcopyrite, BM 88581, from the Greville collection. Bought 1810. This is a flat solid mass of brassy chalcopyrite without matrix 3 x 2 x 1/4 inches with a label saying 'Alva, Stirlingshire'.

Erythrite, Allan-Greg collection AG5. BM 96848 purchased 1860. The description on the label says 'pale pink powdery with annabergite and chloanthite'. It is 1 1/2 x 1 x 1 inches powdery pale pink with patches of a massive grey metallic mineral.

Russell collection. Erythrite originally from the Williams collection, Scorrier House, Cornwall. With an old label 'No. 1 from Alva near Alloa, Clackmannanshire, Scotland'. 4 x 3 x 2 inches pink to reddish masses in and on calcite/barite matrix.

Russell collection. Erythrite from Alva, 'Fife' [*sic*]. Originally from the Lady Elizabeth Anne Coxe Hippisly collection. 3 x 2 x 1/2 inches, heavy mass covered in erythrite with blackish (asbolane?) patches.
Russell collection. Erythrite originally from the Williams collection. Small box of decomposed pieces, the largest one inch, very raspberry red in colour.

31 H. Hutchison: 'The failure of God's Blessing', *Scottish Field* (April 1973).
H. Aitken: 'The Hilderstone silver mine, near Linlithgow', *Trans. Fed. Inst. Min. Eng.* 6 (1893), pp. 193-8.
D. Stephenson, N. J. Fortey and M. J. Gallagher: 'Polymetallic Mineralisation in Carboniferous Rocks at Hilderston near Bathgate, Central Scotland', *British Geological Survey, Mineral Reconnaissance Program Report,* no. 68.
T. K. Meikle: 'Native silver from Hilderston mine, West Lothian, Scotland', *Journal of the Russell Society* 5 (1994), pp. 83-90.

32 R. Z. Jassim: 'Lithogeochemical and mineralogical studies of the silver-copper-baryte deposits of the Ochil Hills, Midland Valley of Scotland' (M.Sc. Thesis, University of Strathclyde 1979). The abstract of this thesis is published in *Trans. Inst. Mining Metal* (1981), p. B91.

33 R. Z. Jassim, R. A. D. Pattrick and M. J. Russell: 'On the origin of the silver + copper + cobalt + baryte mineralization of Ochil Hills, Scotland: a sulphur isotope study', *Trans. Inst. Mining Metal* 92 (1983), pp. B213-B216.

34 J. G. Harrison: 'Heavy metal mines in the Ochil Hills: Chronology and Context', *Forth Naturalist and Historian* 23 (2003), pp. 105-17.

35 J. S. Coats, M. H. Shaw, M. J. Gallagher, M. Armstrong, P. G. Greenwood, B. C. Chacksfield, J. P. Williamson and N. J. Fortey: 'Gold in the Ochil Hills, Scotland', *British Geological Survey, Mineral Reconnaissance Program Report* no. 116 (1991).